Chapter 1: Introduction to SSRIs and Their Role in Mental Health

In the realm of modern medicine, the treatment of mental health disorders has seen substantial advancements over the past few decades. One of the most significant breakthroughs in psychiatric care is the development of serotonin reuptake inhibitors, commonly known as SSRIs. These medications have revolutionized the management of conditions such as depression, anxiety, obsessive-compulsive disorder (OCD), panic disorders, and other mood-related disorders. In this chapter, we will explore the significance of SSRIs in the field of mental health, focusing on how they work, their uses, and the essential role serotonin plays in emotional well-being.

Overview of SSRIs and Their Significance in Modern Medicine

Serotonin reuptake inhibitors are a class of drugs primarily used to increase serotonin levels in the brain, which helps regulate mood, anxiety, and stress levels. By preventing the reabsorption, or reuptake, of serotonin back into the nerve cells, SSRIs increase the availability of this neurotransmitter in the brain. This increase in serotonin availability is believed to play a central role in improving symptoms of mental health disorders.

SSRIs are considered first-line treatment for many mental health conditions, especially depression and anxiety. They have gained popularity due to their relatively favorable side effect profile compared to older classes of antidepressants, such as tricyclic antidepressants (TCAs) and monoamine oxidase inhibitors (MAOIs). These drugs have been proven to improve the quality of life for millions of people, offering hope for individuals struggling with mental health challenges.

The introduction of SSRIs marked a shift from traditional methods of treating mental health conditions, such as therapy or the use of medications with harsher side effects, to a more scientifically targeted approach. With their ability to manage mood disorders effectively, SSRIs have become one of the most widely prescribed classes of medications worldwide.

The Connection Between Serotonin and Mental Health

Serotonin is a neurotransmitter, a chemical messenger that transmits signals in the brain. It is often referred to as the "feel-good" neurotransmitter because it plays a pivotal role in regulating mood, appetite, digestion, and sleep. When serotonin levels are balanced, individuals tend to experience a sense of well-being, calmness, and emotional stability. On the other hand, a deficiency or imbalance in serotonin is often linked to various mental health conditions, particularly those that involve mood regulation, such as depression and anxiety disorders.

The connection between serotonin and mental health is so profound that imbalances in this neurotransmitter are considered one of the key contributors to conditions such as:

- **Depression**: Low levels of serotonin are often associated with feelings of sadness, hopelessness, and a lack of motivation, which are hallmarks of depression.
- **Anxiety**: Serotonin also plays a role in regulating stress responses. Inadequate serotonin may contribute to heightened feelings of worry and anxiety.
- **Obsessive-Compulsive Disorder (OCD)**: Abnormal serotonin functioning has been linked to the intrusive thoughts and compulsive behaviors characteristic of OCD.
- **Panic Disorder**: Reduced serotonin availability can contribute to the panic attacks and heightened sense of fear associated with panic disorder.

Because serotonin affects such a wide range of bodily and emotional functions, regulating its levels in the brain is a key component in treating mood and anxiety disorders. SSRIs are specifically designed to target this neurotransmitter, offering a therapeutic approach that enhances serotonin's effects on the brain's mood-regulating centers.

The Importance of Serotonin Regulation in Emotional Well-Being

Emotional well-being is intricately tied to how the brain processes and regulates neurotransmitters like serotonin. When serotonin levels are balanced, individuals typically experience emotional stability and a positive outlook on life. On the contrary, when serotonin is deficient or poorly regulated, people may experience symptoms of depression, anxiety, and other related conditions.

Regulation of serotonin is crucial not just for alleviating symptoms of mental illness but for maintaining overall mental health. This is why SSRIs are not just effective for individuals already diagnosed with mood disorders; they also play a preventive role in maintaining emotional balance and reducing the risk of relapse.

For individuals living with chronic conditions such as major depressive disorder or generalized anxiety disorder, SSRIs provide long-term benefits by sustaining serotonin levels, thereby preventing the emotional and cognitive disruptions that can be caused by serotonin deficiencies. This helps to improve emotional resilience, reduce symptoms, and enhance overall mental wellness.

However, achieving balanced serotonin levels does not only rely on medication. SSRIs work best when combined with lifestyle factors such as a healthy diet, exercise, and adequate sleep, all of which support serotonin production and overall mental health. Therefore, SSRIs represent one crucial element in a broader, holistic approach to mental health care.

In summary, SSRIs have become a cornerstone of modern psychiatric treatment due to their effectiveness in managing mood disorders and their ability to regulate serotonin levels in the brain. Understanding how serotonin affects mental health and how SSRIs work to balance this neurotransmitter is essential for appreciating their role in promoting emotional well-being and improving the quality of life for millions of individuals.

As we move through the chapters of this book, we will delve deeper into the workings of SSRIs, their specific uses, safety considerations, side effects, and how these medications can be used effectively in treating various mental health conditions.

Chapter 2: Understanding Serotonin and Its Impact on the Body

Serotonin, often referred to as the "feel-good" neurotransmitter, plays a critical role in regulating various functions within the body, most notably in the brain. This chapter delves into the complex nature of serotonin, its multifaceted role in the body, and how it directly influences mood, appetite, sleep, digestion, and overall mental well-being. Understanding serotonin's impact on the body is essential to grasp the importance of serotonin reuptake inhibitors (SSRIs) in treating mood disorders.

The Role of Serotonin as a Neurotransmitter

Serotonin, scientifically known as 5-hydroxytryptamine (5-HT), is a neurotransmitter—one of the many chemical messengers that transmit signals throughout the brain and nervous system. Produced primarily in the brain and intestines, serotonin is synthesized from the amino acid tryptophan, which is obtained through the diet. Once synthesized, serotonin is released by nerve cells and acts on target receptors in the brain and various parts of the body.

Serotonin is involved in a wide range of bodily functions beyond just mood regulation. It plays an essential role in regulating appetite, sleep, digestion, and even cardiovascular function. However, its most well-known and widely studied function is its impact on mood and emotional regulation. In fact, serotonin is often called the "happiness hormone" due to its significant influence on feelings of well-being and contentment.

In the brain, serotonin acts primarily in the central nervous system (CNS), influencing various regions responsible for controlling emotions, cognition, and social behavior. For example, serotonin interacts with areas like the prefrontal cortex, hippocampus, and amygdala, regions known to be involved in mood regulation, memory, and emotional processing.

How Serotonin Influences Mood, Appetite, Sleep, and Digestion
Mood Regulation

One of the most studied roles of serotonin is its involvement in mood regulation. When serotonin levels are balanced, it helps to maintain an emotional state of calm and well-being. A deficiency in serotonin, however, is commonly associated with mood disorders such as depression and anxiety. Individuals with low serotonin levels often experience feelings of sadness, irritability, or anxiety, making serotonin levels an important factor in mental health.

Many antidepressant medications, including SSRIs, target serotonin levels because of its essential role in regulating mood. By increasing the availability of serotonin in the brain, SSRIs help to alleviate the symptoms of depression and anxiety, making serotonin a central focus in the treatment of these conditions.

Appetite Regulation

Serotonin also plays a key role in controlling appetite and food intake. In the brain, serotonin influences the hypothalamus, which regulates hunger signals. High serotonin levels are often associated with reduced appetite and feelings of satiety, whereas low serotonin levels can lead to increased hunger and cravings, particularly for carbohydrate-rich foods. This is why some people experience weight fluctuations when serotonin levels are out of balance.

For individuals with eating disorders such as bulimia or anorexia, serotonin dysregulation may contribute to abnormal eating patterns and distorted body image. SSRIs, by boosting serotonin levels, can help to regulate appetite and improve eating behavior in these individuals.

Sleep Regulation

Another essential function of serotonin is its role in regulating sleep. Serotonin contributes to the regulation of the sleep-wake cycle, influencing both the onset of sleep and the quality of sleep. Specifically, serotonin affects the production of melatonin, a hormone that signals the body to prepare for sleep.

Serotonin's impact on sleep extends to its involvement in sleep disorders. For example, individuals with depression or anxiety often struggle with insomnia, and one potential explanation for this is serotonin imbalances. By increasing serotonin levels, SSRIs help to stabilize sleep patterns, promoting a more restful and consistent sleep cycle. This improvement in sleep quality can, in turn, have positive effects on mood and overall mental health.

Digestion and Gut Health

Interestingly, serotonin is not only found in the brain but also in the gastrointestinal tract, where it plays a vital role in digestion and gut function. In fact, about 90-95% of the body's serotonin is located in the gut. Serotonin helps regulate the movement of the intestines, ensuring proper digestion and absorption of nutrients. It also plays a role in controlling nausea and the sensation of fullness.

An imbalance in serotonin levels within the gut can lead to gastrointestinal issues such as irritable bowel syndrome (IBS), bloating, and discomfort. Since serotonin levels impact both the brain and the gut, it's no surprise that researchers have found connections between mood disorders and gastrointestinal health. This is another reason why SSRIs are often prescribed to individuals with both mood disorders and digestive issues.

The Connection Between Serotonin and Mental Disorders

The link between serotonin and mental health is well-established in the scientific community. Numerous studies have demonstrated that serotonin plays a pivotal role in mood disorders such as depression, anxiety, obsessive-compulsive disorder (OCD), and even bipolar disorder. Low levels of serotonin are commonly associated with symptoms of these conditions, leading to the widespread use of SSRIs and other medications aimed at increasing serotonin availability.

Depression and Anxiety

In conditions like depression and anxiety, serotonin deficiencies contribute to a disrupted mood and emotional regulation. By restoring serotonin levels, SSRIs help to improve mood, reduce feelings of hopelessness, and alleviate anxiety symptoms. For individuals struggling with depression, restoring serotonin balance is a key part of the healing process, which is why SSRIs are often prescribed as the first-line treatment.

Obsessive-Compulsive Disorder (OCD)

OCD is another condition in which serotonin plays a key role. In OCD, serotonin dysregulation is thought to contribute to the repetitive behaviors and intrusive thoughts that characterize the disorder. SSRIs, by increasing serotonin levels, have been shown to help reduce the severity of these symptoms, making them an effective treatment option for those with OCD.

Eating Disorders and Impulse Control

Serotonin also affects behaviors related to eating and impulse control. Conditions like binge eating, bulimia, and anorexia are often associated with serotonin imbalances. By restoring serotonin function, SSRIs can help reduce the compulsive behaviors associated with these disorders and promote healthier eating patterns.

Conclusion

Serotonin is a critical neurotransmitter that influences many essential bodily functions, including mood regulation, appetite control, sleep, digestion, and even gastrointestinal health. Its profound impact on mental well-being has made it a focal point in the treatment of various mental health conditions. By understanding serotonin's role in the body, we can appreciate the importance of serotonin reuptake inhibitors (SSRIs) in restoring balance to individuals suffering from mood disorders, anxiety, OCD, and more. This understanding also highlights the importance of maintaining optimal serotonin levels for overall emotional health and wellness.

In the next chapter, we will delve into the mechanism of SSRIs, exploring how these medications work at the molecular level to increase serotonin availability and help alleviate the symptoms of mental health disorders.

Chapter 3: The Mechanism of SSRIs

Selective Serotonin Reuptake Inhibitors, or SSRIs, are a class of medications designed to manage mood disorders by influencing serotonin levels in the brain. To understand how SSRIs work, it's essential to first understand the basic process of neurotransmission and how serotonin is involved in the regulation of mood and behavior. In this chapter, we will explore the molecular workings of SSRIs, focusing on their mechanism of action, how they inhibit serotonin reuptake, and the resulting increase in serotonin availability in the brain.

How SSRIs Work at the Molecular Level

At the core of the mechanism of SSRIs is their ability to target and interact with serotonin transporters. These transporters are proteins located on the surface of neurons in the brain that are responsible for the reuptake (or reabsorption) of serotonin after it has been released into the synaptic cleft—the small gap between two nerve cells. The reuptake process removes serotonin from the synaptic cleft and returns it to the presynaptic neuron, which reduces serotonin availability and the effectiveness of serotonin signaling.

SSRIs work by specifically inhibiting these serotonin transporters. When an SSRI is introduced into the bloodstream and reaches the brain, it binds to the serotonin transporter molecules. This binding effectively blocks the transporter's ability to reabsorb serotonin, preventing the neurotransmitter from being cleared from the synaptic cleft. As a result, serotonin remains in the synaptic space for a longer period, allowing for increased signaling between neurons. This prolonged serotonin activity enhances the neurotransmitter's effects, leading to mood regulation, reduction of anxiety, and improvement in other mental health symptoms.

The inhibitory action of SSRIs on serotonin reuptake occurs selectively, meaning that SSRIs primarily affect serotonin rather than other neurotransmitters such as dopamine or norepinephrine. This selectivity is one of the key features that distinguishes SSRIs from other classes of antidepressants, such as tricyclic antidepressants (TCAs) and monoamine oxidase inhibitors (MAOIs), which have a broader impact on multiple neurotransmitter systems.

Inhibition of Serotonin Reuptake and Its Effects

The inhibition of serotonin reuptake by SSRIs results in an increase in serotonin levels in the synaptic cleft. This effect is critical because serotonin has widespread influence on brain function. Increased serotonin availability enhances communication between neurons involved in mood regulation, cognition, sleep, and anxiety control.

One of the key brain regions affected by serotonin is the **limbic system**, which plays a central role in emotional processing, mood regulation, and stress responses. In conditions such as depression and anxiety, serotonin levels are often found to be lower than normal, leading to mood disturbances and increased sensitivity to stress. By inhibiting serotonin reuptake, SSRIs help restore the balance of serotonin in these key brain regions, alleviating symptoms associated with depression, anxiety, and other mood disorders.

Furthermore, serotonin has a well-established role in regulating the **hypothalamus**, which controls appetite, sleep, and bodily temperature. The increased serotonin activity promoted by SSRIs can also help to stabilize sleep patterns, reduce excessive appetite, and regulate other functions that may be disrupted in individuals with mood disorders.

The Increase in Serotonin Availability in the Brain

The main therapeutic benefit of SSRIs is the increase in serotonin availability in the brain, which leads to several important effects. This increase is typically gradual, as it takes time for SSRIs to accumulate in the brain and reach therapeutic levels.

1. **Mood Improvement**: Increased serotonin activity helps to stabilize mood, reduce feelings of sadness, hopelessness, and irritability, which are hallmarks of depression. Over time, as serotonin levels increase, individuals may experience a noticeable improvement in their mood and outlook on life.

2. **Reduction of Anxiety**: In addition to its role in mood regulation, serotonin is involved in controlling anxiety responses. By enhancing serotonin activity in areas of the brain associated with fear and stress responses, SSRIs can help reduce the symptoms of anxiety disorders, such as excessive worry, restlessness, and panic attacks.

3. **Improvement in Cognitive Function**: Serotonin has a significant impact on cognitive functions, including memory, attention, and decision-making. By increasing serotonin availability, SSRIs can help improve cognitive clarity and reduce the mental fog that often accompanies depression or anxiety.

4. **Physical Symptoms**: In many individuals, mood disorders are accompanied by physical symptoms such as disrupted sleep, fatigue, and appetite changes. By promoting serotonin activity, SSRIs can help to regulate sleep patterns, increase energy levels, and restore appetite balance, improving overall well-being.

While the increase in serotonin availability may result in these therapeutic benefits, it is important to note that the full effects of SSRIs may not be felt immediately. It typically takes 2 to 4 weeks for individuals to experience the full benefit of SSRI treatment. During this time, the brain's serotonin system gradually adjusts to the increased levels of serotonin, leading to an improvement in mood, energy, and overall function.

Long–Term Effects of SSRIs on Serotonin Regulation

SSRIs not only help to correct serotonin imbalances in the short term, but they may also have lasting effects on serotonin regulation in the brain. Over time, the brain may become more responsive to serotonin, and the receptor systems that interact with serotonin may undergo changes. These adaptive changes are thought to contribute to the long-term benefits of SSRIs, helping to prevent relapse and maintain emotional balance.

For individuals undergoing long-term SSRI treatment, these changes in serotonin system functioning are important for promoting sustained recovery from depression, anxiety, and other mood disorders. By enhancing serotonin transmission and creating a more stable serotonin environment, SSRIs contribute to long-term emotional resilience and mental well-being.

Conclusion

SSRIs are a class of medications that work by selectively inhibiting the serotonin transporter, preventing the reabsorption of serotonin and allowing for increased serotonin availability in the brain. This increase in serotonin helps to regulate mood, reduce anxiety, improve cognitive function, and stabilize sleep patterns, all of which are essential for mental health. Understanding the molecular mechanism of SSRIs and how they influence serotonin activity is crucial for appreciating their role in treating mood disorders and enhancing emotional well-being.

In the following chapters, we will explore how SSRIs are used in the treatment of various mental health conditions, their safety profile, potential side effects, and practical guidelines for their use in long-term wellness.

Chapter 4: A History of SSRIs

The development of selective serotonin reuptake inhibitors (SSRIs) represents a significant milestone in the treatment of mood disorders, marking a shift in how mental health conditions have been approached over the past several decades. Understanding the history of SSRIs offers insight into their transformative role in modern psychiatric care and highlights how these medications became a cornerstone of antidepressant therapy. In this chapter, we will explore the development of SSRIs, their emergence in the context of antidepressant therapy, their impact on the landscape of mental health treatment, and how they compare to older classes of antidepressants.

Development of SSRIs in the Context of Antidepressant Therapy

The journey toward the creation of SSRIs began with the discovery and exploration of serotonin's role in the brain. As early as the 1950s and 1960s, scientists had already identified serotonin as a key neurotransmitter involved in regulating mood, anxiety, and other emotional functions. However, it wasn't until the 1970s and 1980s that researchers began to understand the connection between serotonin and psychiatric conditions such as depression and anxiety.

Initially, antidepressant therapies focused on the use of **monoamine oxidase inhibitors** (MAOIs) and **tricyclic antidepressants** (TCAs), which were among the first widely used medications for treating mood disorders. While these medications were effective for many patients, they came with a range of undesirable side effects. MAOIs, for instance, required strict dietary restrictions to avoid dangerous interactions with foods containing tyramine, and TCAs were known for causing sedation, weight gain, and cardiovascular issues. These side effects significantly limited the broad use of these medications.

In the search for a safer and more effective treatment option, researchers focused on serotonin as a key target. The idea was simple: if boosting serotonin levels in the brain could help alleviate symptoms of depression, then developing a drug that could increase serotonin availability by blocking its reuptake might offer a more targeted, effective, and safer solution. This vision eventually led to the creation of SSRIs.

The first SSRI to be developed was **fluoxetine**, which was introduced in 1987 under the brand name **Prozac**. Fluoxetine's selective action on serotonin reuptake—without significantly affecting other neurotransmitters like norepinephrine and dopamine—offered a major advancement over previous antidepressants. Unlike TCAs and MAOIs, Prozac did not carry the same risk of serious side effects, and its more targeted mechanism of action made it a breakthrough in antidepressant therapy.

How SSRIs Changed the Landscape of Mental Health Treatment

The introduction of SSRIs fundamentally transformed the landscape of mental health treatment. For decades, patients suffering from depression and anxiety had limited treatment options, many of which were associated with troublesome side effects and complex dosing regimens. SSRIs changed that by offering a medication that was easier to use, with a more favorable side-effect profile, and with the ability to be prescribed for longer periods of time without the need for constant monitoring.

SSRIs not only revolutionized the treatment of depression, but they also offered new hope for individuals suffering from anxiety disorders, obsessive-compulsive disorder (OCD), panic disorders, and post-traumatic stress disorder (PTSD). SSRIs provided a reliable and relatively safe way to manage these often debilitating conditions, allowing people to experience improved quality of life with fewer disruptions to their daily routines.

The widespread acceptance of SSRIs among healthcare providers also helped reduce the stigma surrounding the treatment of mental health conditions. SSRIs, especially Prozac, became culturally significant in the 1990s, widely regarded as a symbol of the growing recognition of mental health issues as legitimate medical conditions. The availability of these medications helped to normalize the treatment of depression and anxiety, encouraging more people to seek help for their mental health struggles without the fear of being labeled as "weak" or "unstable."

Comparison with Older Classes of Antidepressants

SSRIs are often compared to older classes of antidepressants, such as **tricyclic antidepressants (TCAs)** and **monoamine oxidase inhibitors (MAOIs)**, because these medications were among the first widely used antidepressants. While all three classes of drugs share the goal of improving mood by increasing the levels of neurotransmitters in the brain, there are significant differences in their mechanisms, side effects, and overall safety profiles.

Tricyclic Antidepressants (TCAs)

TCAs, such as **amitriptyline** and **nortriptyline**, were introduced in the 1950s and became one of the most commonly prescribed antidepressants for several decades. TCAs work by inhibiting the reuptake of both serotonin and norepinephrine, but they also affect a range of other receptors in the body, which leads to their broad array of side effects. Common side effects of TCAs include weight gain, sedation, dry mouth, blurred vision, and orthostatic hypotension (a drop in blood pressure when standing). Additionally, TCAs are known to be cardiotoxic at high doses, making them particularly dangerous in cases of overdose.

In contrast, SSRIs have a much more targeted action on serotonin, leading to fewer side effects. While SSRIs still have their own set of potential side effects (discussed in Chapter 8), they are generally considered safer, particularly in terms of overdose risk. SSRIs also tend to be better tolerated over the long term, which makes them a preferable option for many patients.

Monoamine Oxidase Inhibitors (MAOIs)

MAOIs, such as **phenelzine** and **tranylcypromine**, work by inhibiting the enzyme monoamine oxidase, which breaks down serotonin, norepinephrine, and dopamine in the brain. By inhibiting this enzyme, MAOIs increase the levels of these neurotransmitters, which can improve mood and alleviate symptoms of depression.

However, MAOIs come with significant risks. One of the most dangerous side effects of MAOIs is the potential for **hypertensive crises**, which can occur if a patient consumes foods or drinks containing high levels of **tyramine** (such as aged cheeses, cured meats, and certain alcoholic beverages). Tyramine is normally broken down by monoamine oxidase, but with this enzyme inhibited, tyramine can build up to dangerous levels, leading to dangerously high blood pressure. As a result, patients on MAOIs must adhere to strict dietary restrictions, which can be inconvenient and difficult to maintain.

SSRIs, by contrast, do not have these dietary restrictions, and their side-effect profile is much more manageable. This convenience, combined with their effectiveness, has made SSRIs the preferred treatment option for most patients with depression and anxiety disorders.

Conclusion

The development of SSRIs was a turning point in the treatment of mental health conditions. From their first introduction with Prozac in 1987, SSRIs have significantly improved the management of mood disorders, anxiety, and other psychiatric conditions. By providing a safer, more targeted alternative to older antidepressants like TCAs and MAOIs, SSRIs have not only changed the way mental health conditions are treated but have also helped to reduce the stigma surrounding mental health treatment. Their impact on both the scientific community and the general public has been profound, and they continue to serve as a cornerstone in the pharmacological treatment of mental health conditions.

In the next chapter, we will explore the primary uses of SSRIs, focusing on their effectiveness in treating common conditions such as depression, anxiety disorders, obsessive-compulsive disorder, and more.

Chapter 6: Exploring Common SSRIs

Selective serotonin reuptake inhibitors (SSRIs) are a group of antidepressants that target the serotonin system in the brain. While there are several SSRIs on the market, each medication has unique characteristics and is used for different therapeutic purposes. Understanding the individual profiles of commonly prescribed SSRIs helps provide insight into their effectiveness, side effects, and how they can be tailored to individual patient needs. In this chapter, we will explore some of the most widely prescribed SSRIs, including **fluoxetine**, **sertraline**, **paroxetine**, **citalopram**, and **escitalopram**, highlighting their specific attributes, uses, and differences.

Fluoxetine (Prozac)

Fluoxetine, one of the most widely recognized SSRIs, was the first SSRI to be approved by the U.S. Food and Drug Administration (FDA) in 1987 under the brand name **Prozac**. It quickly became a go-to treatment for depression, largely due to its relatively mild side effect profile compared to older antidepressants.

Therapeutic Uses:

- **Depression**: Fluoxetine is primarily used to treat major depressive disorder (MDD). It is often prescribed as a first-line treatment due to its effectiveness and tolerability.

- **Anxiety Disorders**: Fluoxetine is also effective in treating generalized anxiety disorder (GAD), panic disorder, and social anxiety disorder.

- **Obsessive-Compulsive Disorder (OCD)**: Fluoxetine has been shown to be effective in reducing the symptoms of OCD, including intrusive thoughts and compulsive behaviors.

- **Bulimia Nervosa**: Fluoxetine is often prescribed as part of the treatment for eating disorders like bulimia.

Side Effects:

While generally well-tolerated, fluoxetine can cause some side effects, including:

- Insomnia

- Nervousness or anxiety

- Nausea

- Decreased libido

- Weight loss (common in the initial stages of treatment)

Fluoxetine is known for its relatively long half-life, which means it stays in the body for a longer period. This can be advantageous because it allows for a more gradual reduction of the medication if needed, but it may also lead to a slower onset of therapeutic effects.

Sertraline (Zoloft)

Sertraline, marketed under the brand name **Zoloft**, was approved by the FDA in 1991 and has become one of the most prescribed SSRIs worldwide. It shares many uses with fluoxetine but is known for its slightly different side-effect profile.

Therapeutic Uses:

- **Depression**: Sertraline is commonly prescribed for patients with major depressive disorder and has proven to be effective for many individuals.
- **Anxiety Disorders**: It is particularly effective in treating anxiety disorders such as generalized anxiety disorder (GAD), panic disorder, and social anxiety disorder.
- **Obsessive-Compulsive Disorder (OCD)**: Like fluoxetine, sertraline is also prescribed for OCD treatment.
- **Post-Traumatic Stress Disorder (PTSD)**: Sertraline has been shown to help alleviate symptoms of PTSD, such as flashbacks, hypervigilance, and emotional numbness.

Side Effects:

Common side effects include:

- Sexual dysfunction (such as decreased libido or delayed orgasm)
- Insomnia
- Weight gain (although this is less common compared to other SSRIs)
- Fatigue
- Nausea or gastrointestinal discomfort

Sertraline is generally well-tolerated, but its potential to cause sexual side effects is significant. This can be a major concern for some patients, and managing these side effects often requires close communication with healthcare providers.

Paroxetine (Paxil)

Paroxetine, sold under the brand name **Paxil**, is another widely prescribed SSRI that was first approved by the FDA in 1992. Paroxetine has a more sedating effect compared to other SSRIs, which can be beneficial for patients with anxiety but may not be ideal for those who experience fatigue or daytime drowsiness.

Therapeutic Uses:

- **Depression**: Paroxetine is effective in treating major depressive disorder and has been shown to improve mood and reduce symptoms of depression.

- **Anxiety and Panic Disorders**: It is particularly useful for patients with generalized anxiety disorder (GAD), social anxiety disorder, and panic disorder.

- **Post-Traumatic Stress Disorder (PTSD)**: Paroxetine is one of the first SSRIs to be approved by the FDA for treating PTSD.

- **Premenstrual Dysphoric Disorder (PMDD)**: Paroxetine can help alleviate severe symptoms of PMDD, which are similar to premenstrual syndrome (PMS) but more intense.

Side Effects:

While effective, paroxetine is associated with certain side effects, including:

- Weight gain (more pronounced compared to other SSRIs)

- Sedation or drowsiness (which may be useful for anxious patients, but less ideal for those needing alertness)

- Sexual dysfunction (common across most SSRIs)

- Dry mouth

- Increased sweating

Due to its sedating nature, paroxetine may be more suitable for patients dealing with anxiety or those who experience difficulty sleeping. However, its potential for weight gain and sedation may be problematic for some individuals.

Citalopram (Celexa)

Citalopram, sold under the brand name **Celexa**, is one of the older SSRIs approved in 1998. It is known for its relatively straightforward side effect profile and mild sedation, making it an effective and well-tolerated treatment for depression and anxiety disorders.

Therapeutic Uses:

- **Depression**: Citalopram is commonly prescribed for major depressive disorder.
- **Anxiety**: It is also used for generalized anxiety disorder (GAD) and other anxiety-related conditions.
- **Off-label Uses**: Citalopram may be used for other conditions such as panic disorder and OCD, though it is not as commonly prescribed for these as some other SSRIs.

Side Effects:

Citalopram is generally well-tolerated, but it can cause:

- Insomnia or sleep disturbances

- Sexual dysfunction

- Nausea

- Dizziness or lightheadedness

One potential concern with citalopram, particularly at higher doses, is its potential to cause **QT prolongation**, a heart rhythm abnormality that can be dangerous. As such, higher doses are typically avoided, particularly in patients with pre-existing heart conditions.

Escitalopram (Lexapro)

Escitalopram, marketed as **Lexapro**, is a more refined version of citalopram, approved by the FDA in 2002. The main difference between escitalopram and citalopram is that escitalopram is the S-enantiomer of citalopram, which means it is chemically more active and may offer a more potent and targeted effect with fewer side effects.

Therapeutic Uses:

- **Depression**: Escitalopram is highly effective for major depressive disorder.

- **Anxiety Disorders**: It is one of the most commonly prescribed SSRIs for generalized anxiety disorder and other anxiety-related conditions.

- **Off-label Uses**: Escitalopram is also effective in treating panic disorder and OCD.

Side Effects:

The side effect profile of escitalopram is similar to that of other SSRIs but is often better tolerated due to its more targeted action. Common side effects include:

- Sexual dysfunction
- Insomnia
- Nausea
- Fatigue

Escitalopram has a slightly lower risk of weight gain and sedation compared to other SSRIs, which may make it a better choice for patients concerned about these issues.

Conclusion

Each SSRI has unique characteristics, but they all share a common mechanism of action—selectively inhibiting serotonin reuptake to increase serotonin availability in the brain. The choice of SSRI often depends on the individual patient's medical history, symptoms, and preferences, as well as the specific side-effect profiles of each medication.

Fluoxetine, sertraline, paroxetine, citalopram, and escitalopram are some of the most commonly prescribed SSRIs, and while they are all effective for treating depression and anxiety, their differences in side effects, sedation levels, and interactions with other medications make them suitable for different types of patients. Understanding the profiles of these SSRIs allows healthcare providers to personalize treatment and find the most effective medication with the fewest side effects for each patient.

In the following chapters, we will explore the safety and efficacy of SSRIs, as well as their side effects, to help individuals and healthcare providers make informed decisions regarding SSRI treatment.

Chapter 5: The Primary Uses of SSRIs

Selective serotonin reuptake inhibitors (SSRIs) are widely prescribed for a variety of mental health conditions due to their effectiveness in balancing serotonin levels in the brain. Since their introduction, SSRIs have become one of the most frequently used treatments for disorders such as depression, anxiety, obsessive-compulsive disorder (OCD), panic disorder, eating disorders, and post-traumatic stress disorder (PTSD). This chapter delves into the primary uses of SSRIs, highlighting the conditions they are prescribed for, how they work in the treatment process, and the benefits they provide in improving patients' quality of life.

Treatment of Depression and Anxiety Disorders
Depression

Depression is one of the most common mental health conditions worldwide, affecting millions of individuals across all age groups. It is characterized by persistent feelings of sadness, hopelessness, and a loss of interest in daily activities. Depression can interfere with an individual's ability to function, affecting work, relationships, and overall well-being.

SSRIs are considered one of the first-line treatments for major depressive disorder (MDD) because of their ability to increase serotonin availability in the brain, helping to regulate mood and emotional response. By inhibiting serotonin reuptake, SSRIs help alleviate the symptoms of depression, such as low energy, sadness, and anhedonia (the inability to feel pleasure). SSRIs are particularly favored for depression because they are well-tolerated and cause fewer side effects compared to older classes of antidepressants like tricyclic antidepressants (TCAs) and monoamine oxidase inhibitors (MAOIs).

The treatment of depression with SSRIs typically involves a combination of medication and psychotherapy. Cognitive-behavioral therapy (CBT) is often used alongside SSRIs to help individuals develop healthier thinking patterns and coping strategies.

Anxiety Disorders

Anxiety disorders, including generalized anxiety disorder (GAD), social anxiety disorder (SAD), and panic disorder, are characterized by excessive and often irrational worry, fear, or nervousness. Individuals with these conditions may experience physical symptoms like heart palpitations, sweating, and difficulty breathing, along with emotional distress.

SSRIs are effective in treating various anxiety disorders because serotonin is closely involved in the brain's regulation of fear and stress responses. By increasing serotonin levels, SSRIs help individuals feel calmer and more in control of their emotional responses. SSRIs, such as sertraline (Zoloft) and fluoxetine (Prozac), are particularly effective in treating GAD and panic disorder. They work to reduce the frequency and intensity of anxiety attacks, providing long-term relief for individuals with chronic anxiety.

The benefits of SSRIs in treating anxiety are not immediate, and it often takes several weeks for individuals to notice significant improvement. However, over time, patients typically experience a reduction in overall anxiety, with fewer panic attacks and less social apprehension.

Addressing Obsessive–Compulsive Disorder (OCD)

Obsessive-compulsive disorder (OCD) is a chronic condition characterized by the presence of unwanted, intrusive thoughts (obsessions) and the compulsive behaviors that individuals feel compelled to perform in response to these thoughts. Common compulsions include hand-washing, counting, or repeating certain actions in an attempt to reduce the anxiety caused by the obsessions.

SSRIs are a mainstay in the treatment of OCD due to their ability to regulate serotonin levels in the brain, which play a significant role in the obsessions and compulsions that define the disorder. Fluoxetine (Prozac), sertraline (Zoloft), and fluvoxamine (Luvox) are commonly prescribed for OCD. These medications can help reduce both the frequency of intrusive thoughts and the intensity of compulsive behaviors.

SSRIs may not fully eliminate the symptoms of OCD but are effective in significantly reducing them, making the disorder more manageable. In many cases, SSRIs are used in conjunction with behavioral therapies, particularly exposure and response prevention (ERP), a form of cognitive-behavioral therapy (CBT) that helps individuals gradually confront their fears without engaging in compulsive behaviors.

The Role of SSRIs in Panic Disorders and Eating Disorders
Panic Disorders

Panic disorder is characterized by recurrent and unexpected panic attacks—intense episodes of fear or discomfort that occur without warning. Panic attacks can cause physical symptoms like chest pain, dizziness, and shortness of breath, which may lead individuals to fear future attacks and avoid certain situations.

SSRIs are effective in reducing the frequency and severity of panic attacks by increasing serotonin availability in the brain, which helps to regulate stress responses. Sertraline (Zoloft) and fluoxetine (Prozac) are commonly prescribed for panic disorder, as they help decrease the hypervigilance and anxiety that contribute to panic attacks. Over time, SSRIs help patients feel more confident in facing situations that previously triggered panic attacks, promoting a sense of control and emotional stability.

Eating Disorders

Eating disorders, such as **bulimia nervosa** and **binge-eating disorder**, are characterized by abnormal eating patterns, preoccupation with body image, and a loss of control over food intake. These disorders often coexist with emotional distress, depression, and anxiety, which makes them particularly challenging to treat.

SSRIs are commonly used to treat the underlying emotional and psychological issues associated with eating disorders, such as depression and anxiety. Fluoxetine (Prozac) is the most commonly prescribed SSRI for bulimia, as it has been shown to reduce binge-eating episodes and purging behaviors. SSRIs work by improving mood regulation and reducing the anxiety that often drives disordered eating behaviors.

While SSRIs are not a cure for eating disorders, they can play a critical role in managing the emotional symptoms that fuel the disorders. SSRIs are often used in conjunction with psychotherapy, particularly cognitive-behavioral therapy (CBT), to address both the psychological and behavioral aspects of eating disorders.

SSRIs in the Management of Post–Traumatic Stress Disorder (PTSD)

Post-traumatic stress disorder (PTSD) occurs when individuals experience a traumatic event that causes persistent emotional and psychological distress. Symptoms of PTSD include flashbacks, nightmares, intrusive memories, and heightened anxiety. Individuals with PTSD often experience emotional numbness and difficulty managing relationships, work, and daily tasks.

SSRIs, particularly sertraline (Zoloft) and paroxetine (Paxil), are first-line treatments for PTSD. These medications help to alleviate symptoms of hyperarousal, anxiety, and mood swings by increasing serotonin availability in the brain. Over time, SSRIs help to reduce the frequency of flashbacks and nightmares, allowing individuals to process traumatic memories more effectively. While SSRIs may not eliminate all symptoms of PTSD, they are effective in significantly reducing distress and helping patients regain a sense of control over their lives.

Conclusion

SSRIs are versatile medications with broad applications in the treatment of various mental health conditions. From depression and anxiety to obsessive-compulsive disorder, panic disorder, eating disorders, and post-traumatic stress disorder, SSRIs offer relief from some of the most challenging psychiatric conditions. By targeting serotonin reuptake and enhancing serotonin availability in the brain, SSRIs help regulate mood, reduce anxiety, and improve the overall quality of life for individuals with these disorders.

The versatility and effectiveness of SSRIs make them a cornerstone of modern psychiatric treatment, often used in combination with therapy to provide a holistic approach to mental health care. In the following chapters, we will explore the safety and efficacy of SSRIs, common side effects, and how to optimize their use in long-term treatment for mental health and wellness.

Chapter 7: Safety and Efficacy of SSRIs

Selective serotonin reuptake inhibitors (SSRIs) are widely recognized for their effectiveness in treating a range of mental health conditions, from depression and anxiety to obsessive-compulsive disorder (OCD) and panic attacks. While SSRIs have revolutionized the treatment of mood disorders, it's crucial to examine their safety and efficacy to fully understand their role in long-term mental health care. In this chapter, we will explore the safety profile of SSRIs, the scientific evidence supporting their use, and the considerations for their long-term use in managing mental health conditions.

SSRIs as Generally Safe Treatments

One of the key reasons SSRIs have become a first-line treatment for conditions like depression and anxiety is their favorable safety profile compared to earlier generations of antidepressants, such as tricyclic antidepressants (TCAs) and monoamine oxidase inhibitors (MAOIs). Unlike these older medications, which are associated with more severe side effects and greater toxicity in overdose situations, SSRIs tend to have a milder side effect profile and a lower risk of fatal overdose.

SSRIs specifically target serotonin reuptake in the brain, without significantly affecting other neurotransmitters like dopamine or norepinephrine. This selective mechanism of action contributes to their lower risk of side effects. They do not usually cause the extreme sedative effects or significant weight gain that are common with tricyclics, nor do they require dietary restrictions, as is the case with MAOIs.

That said, SSRIs are not without risks. Side effects are common, though most are generally mild and resolve over time as the body adjusts to the medication. As with any medication, the decision to start SSRIs must be carefully weighed against potential side effects and individual health conditions. Regular monitoring by healthcare providers ensures that any potential issues are identified and addressed promptly.

The Evidence Supporting Their Efficacy in Managing Mental Health Conditions

SSRIs are among the most researched and well-documented medications for mental health treatment. Numerous clinical studies have demonstrated their efficacy in treating depression, anxiety, and other mood disorders. Below, we review the evidence supporting their use for some of the most common conditions treated by SSRIs.

Depression

Clinical trials consistently show that SSRIs are highly effective in treating major depressive disorder (MDD). Studies have found that SSRIs can significantly improve mood, reduce feelings of hopelessness, and enhance overall functioning in individuals with depression. They are often used as a first-line treatment because they are generally well-tolerated and have fewer side effects than other classes of antidepressants. Fluoxetine (Prozac), sertraline (Zoloft), and escitalopram (Lexapro) are among the most studied SSRIs, and research supports their use in both short-term and long-term management of depression.

Anxiety Disorders

SSRIs are also highly effective in the treatment of various anxiety disorders, including generalized anxiety disorder (GAD), social anxiety disorder (SAD), and panic disorder. By increasing serotonin availability in the brain, SSRIs help regulate the body's response to stress and anxiety, reducing the frequency and intensity of panic attacks, intrusive thoughts, and general nervousness. Studies show that SSRIs can significantly reduce the symptoms of anxiety disorders and improve quality of life for individuals experiencing chronic anxiety.

Obsessive-Compulsive Disorder (OCD)

For individuals with OCD, SSRIs offer relief from intrusive thoughts and compulsive behaviors. Research indicates that SSRIs are effective in reducing both the severity of obsessions (unwanted thoughts) and compulsions (repetitive behaviors). Fluoxetine (Prozac), sertraline (Zoloft), and fluvoxamine (Luvox) are commonly prescribed to help manage OCD symptoms. Though not a cure, SSRIs play a crucial role in reducing symptoms and improving the ability to engage in daily activities without being consumed by the disorder.

Panic Disorders and Post-Traumatic Stress Disorder (PTSD)

SSRIs are considered effective in reducing the frequency and severity of panic attacks, as well as improving overall anxiety levels in panic disorder patients. Similarly, SSRIs have been shown to be effective in managing symptoms of post-traumatic stress disorder (PTSD), including flashbacks, hypervigilance, and emotional numbness. Sertraline (Zoloft) and paroxetine (Paxil) are FDA-approved for PTSD treatment, providing critical relief to individuals suffering from the emotional aftermath of trauma.

Addressing Concerns About Long–Term Use

While SSRIs are considered safe and effective for long-term use, there are several factors that must be monitored and managed throughout the treatment period.

Tolerance and Effectiveness Over Time

One of the concerns with long-term SSRI use is the possibility of **tolerance** or **loss of efficacy**. Some patients report that the effects of SSRIs diminish over time, leading to the need for dose adjustments or a switch to a different medication. However, evidence supporting SSRI tolerance is mixed, and many individuals benefit from long-term treatment without experiencing a loss of effectiveness. It is important for individuals taking SSRIs to have regular follow-up appointments with their healthcare provider to assess their response to treatment and make any necessary adjustments.

The Risk of Relapse

For individuals with chronic conditions like depression and anxiety, long-term SSRI use may be necessary to prevent relapse. Stopping SSRIs prematurely can lead to a return of symptoms, and it is not uncommon for patients to experience a recurrence of depression or anxiety if treatment is abruptly discontinued. To minimize this risk, healthcare providers often recommend a gradual tapering-off process when discontinuing SSRIs, in combination with behavioral therapies or other treatments to maintain progress.

Side Effects and Long–Term Health

While SSRIs are generally well-tolerated, there are certain side effects that can become more noticeable with long-term use. Common side effects include sexual dysfunction (e.g., reduced libido or delayed orgasm), weight changes, insomnia, and gastrointestinal issues. In some cases, these side effects may decrease over time, but in others, they may require a dosage adjustment or a change in medication.

Long-term use of SSRIs has also been linked to an increased risk of **bone fractures** and, in rare cases, **bleeding disorders**. Although the risk of bleeding is generally low, individuals taking SSRIs for extended periods should be monitored for any signs of bleeding or unusual bruising. Additionally, some studies suggest that long-term SSRI use may impact cognitive function in elderly patients, although this is still an area of ongoing research.

The Role of Monitoring in Long-Term Use

Given the potential side effects and long-term considerations, it is essential for individuals on SSRIs to maintain regular communication with their healthcare providers. Regular monitoring helps ensure that the medication remains effective, side effects are managed appropriately, and any potential risks are addressed early on.

Healthcare providers may also consider periodic breaks from SSRI treatment in certain cases to assess the patient's current condition and determine if continued medication is necessary. In some cases, psychological therapies such as cognitive-behavioral therapy (CBT) or mindfulness-based therapies may be used as adjuncts to SSRIs, particularly when medication alone is insufficient.

Conclusion

SSRIs are generally considered safe and effective treatments for a wide range of mental health conditions, including depression, anxiety, OCD, panic disorder, and PTSD. The growing body of research supporting their use demonstrates their efficacy in managing symptoms and improving quality of life for millions of individuals worldwide. While SSRIs are well-tolerated by many patients, long-term use requires ongoing monitoring to manage side effects and ensure continued effectiveness. The combination of SSRIs with behavioral therapies offers a comprehensive approach to mental health treatment, enabling individuals to achieve better emotional regulation and resilience in the face of life's challenges.

In the following chapters, we will explore common side effects of SSRIs, how to manage them, and the importance of a holistic treatment approach that integrates medication with lifestyle factors such as diet, exercise, and self-care.

Chapter 8: Side Effects of SSRIs

While selective serotonin reuptake inhibitors (SSRIs) are widely regarded as safe and effective treatments for mental health conditions, like all medications, they come with the potential for side effects. Understanding these side effects, how to manage them, and when they might require an adjustment in treatment is crucial for optimizing the benefits of SSRIs while minimizing any adverse effects. In this chapter, we will explore the most common side effects associated with SSRIs, as well as strategies for managing these effects to improve the overall treatment experience.

Common Side Effects of SSRIs

SSRIs are generally well-tolerated, but they can cause side effects, especially during the first few weeks of treatment as the body adjusts. Some of the most common side effects include:

1. Insomnia and Sleep Disturbances

Many individuals taking SSRIs experience difficulties with sleep, ranging from trouble falling asleep to disrupted or non-restorative sleep. This side effect is more commonly associated with certain SSRIs, like **fluoxetine** (Prozac), which have a more stimulating effect. However, sleep disturbances are not limited to these medications, and other SSRIs can cause drowsiness, which might result in daytime fatigue.

Managing Sleep Issues:

- **Adjusting timing**: Taking the SSRI in the morning may help reduce insomnia, while taking it in the evening may help with daytime drowsiness.

- **Cognitive Behavioral Therapy for Insomnia (CBT-I)**: This type of therapy can help address sleep disturbances by altering negative sleep patterns and behaviors.

- **Sleep hygiene**: Practicing good sleep hygiene, such as maintaining a regular sleep schedule and avoiding caffeine or electronics before bed, can also improve sleep quality.

2. Headaches

Headaches are one of the most common side effects reported by individuals taking SSRIs. These headaches are often mild to moderate and tend to improve after the first few weeks of treatment. However, they can still be bothersome during the adjustment period.

Managing Headaches:

- **Over-the-counter pain relievers**: Non-prescription medications like ibuprofen or acetaminophen can provide relief.

- **Hydration**: Ensuring proper hydration can also help alleviate tension headaches, which are sometimes triggered by dehydration.

- **Gradual dose adjustments**: Starting with a lower dose and gradually increasing it can reduce the likelihood of headaches during the adjustment phase.

3. Sexual Dysfunction

Sexual side effects, such as reduced libido, delayed orgasm, or erectile dysfunction, are common with SSRIs. These effects can be distressing for many individuals and may contribute to noncompliance with the medication.

Managing Sexual Dysfunction:

- **Switching medications**: If sexual dysfunction persists, healthcare providers may suggest switching to a different SSRI with a lower incidence of sexual side effects, such as **escitalopram** (Lexapro) or **sertraline** (Zoloft).
- **Adjunct medications**: In some cases, medications like **bupropion** (Wellbutrin) may be used alongside SSRIs to help mitigate sexual side effects.
- **Timing of dose**: Taking the medication at different times of the day or altering the dose schedule may sometimes alleviate sexual dysfunction.

4. Gastrointestinal Issues

Gastrointestinal (GI) side effects, including nausea, diarrhea, dry mouth, and indigestion, are also commonly reported during the early stages of SSRI treatment. These symptoms generally subside after the body adjusts to the medication.

Managing GI Issues:

- **Taking medication with food**: This can help reduce nausea and upset stomach. Avoiding large meals or spicy foods during the early stages of treatment may also help.
- **Over-the-counter remedies**: Antacids or anti-nausea medications, under the guidance of a healthcare provider, can help alleviate discomfort.
- **Hydration**: Drinking plenty of fluids and staying hydrated can reduce some gastrointestinal symptoms like dry mouth and constipation.

5. Weight Changes

Weight gain or loss is another side effect associated with SSRIs. While some people experience an increase in appetite and weight gain, others may notice a loss of appetite or weight loss, especially early in treatment. The weight-related side effects are more common with certain SSRIs, such as **paroxetine** (Paxil), which may have more of an appetite-stimulating effect.

Managing Weight Changes:

- **Diet and exercise**: Maintaining a healthy diet and regular exercise routine can help manage weight changes while on SSRIs.

- **Monitoring**: Regular monitoring of weight during treatment can help identify any significant changes early, allowing for timely adjustments in medication or lifestyle.

- **Switching medications**: If weight gain becomes a significant concern, a healthcare provider may recommend a different SSRI with a lower risk of weight changes.

6. Dizziness and Lightheadedness

Some individuals report dizziness or lightheadedness, particularly when standing up quickly. This side effect is often related to the medication's impact on serotonin receptors, which can affect blood pressure regulation.

Managing Dizziness:

- **Getting up slowly**: Taking care to rise gradually from a sitting or lying position can help reduce the risk of dizziness.

- **Staying hydrated**: Dehydration can exacerbate dizziness, so drinking plenty of fluids may help alleviate symptoms.

- **Consulting a healthcare provider**: If dizziness is persistent or severe, a healthcare provider may adjust the dosage or switch to a different medication.

When Side Effects May Require Adjustment of Treatment

While most side effects are temporary and improve over time, some may require changes in treatment. If side effects are particularly bothersome, last for a prolonged period, or significantly interfere with daily life, it is important to consult with a healthcare provider. Here are a few situations where adjustments might be necessary:

- **Unmanageable side effects**: If the side effects are severe or do not subside after a few weeks, switching to a different SSRI or changing to a medication from a different class (e.g., **bupropion** or **mirtazapine**) may be considered.
- **Non-responsiveness**: If the SSRI does not effectively treat the symptoms of depression or anxiety after several weeks, a dosage adjustment or a change in medication may be necessary.
- **Comorbidity concerns**: If the patient has other health conditions, such as heart disease or liver problems, the medication may need to be adjusted to avoid complications.

Conclusion

SSRIs are generally well-tolerated and highly effective in the treatment of various mental health disorders, but like all medications, they come with the potential for side effects. The most common side effects, including insomnia, headaches, sexual dysfunction, and gastrointestinal issues, are typically mild and tend to resolve after the body adjusts to the medication. However, it is essential to monitor side effects and work closely with a healthcare provider to manage any adverse effects. In some cases, adjusting the dosage, switching medications, or adding adjunctive treatments can help alleviate bothersome side effects while still maintaining the therapeutic benefits of the SSRI.

In the next chapter, we will explore a potentially more serious concern when using SSRIs: **serotonin syndrome**—a rare but life-threatening condition that can occur due to excessive serotonin in the brain. Understanding its causes, symptoms, and prevention is vital for ensuring the safe use of SSRIs in the treatment of mental health conditions.

Chapter 9: The Risk of Serotonin Syndrome

Serotonin syndrome is a rare but potentially life-threatening condition that can occur when there is an excessive accumulation of serotonin in the brain. It is a medical emergency that requires immediate attention and treatment. Although serotonin syndrome is most commonly associated with the use of drugs that increase serotonin levels, such as SSRIs, it can also result from the interaction between medications or the use of multiple serotonin-enhancing substances. In this chapter, we will explore serotonin syndrome in detail, including its causes, symptoms, how it is diagnosed, and how it is treated and prevented.

Understanding Serotonin Syndrome and Its Causes

Serotonin syndrome occurs when there is an excess of serotonin in the central nervous system, leading to overstimulation of serotonin receptors in the brain. While serotonin plays a crucial role in regulating mood, appetite, and sleep, too much of it can result in toxic effects that disrupt normal brain and bodily functions.

SSRIs are generally safe when taken as prescribed, but because they increase serotonin levels by inhibiting its reuptake, they have the potential to contribute to serotonin syndrome, particularly when used in combination with other medications or substances that also increase serotonin levels. Some of the most common drugs and substances that can increase the risk of serotonin syndrome include:

- **Other serotonergic drugs**: Taking multiple medications that increase serotonin levels, such as other SSRIs, **serotonin-norepinephrine reuptake inhibitors (SNRIs)**, **monoamine oxidase inhibitors (MAOIs)**, or **tricyclic antidepressants (TCAs)**, can increase the risk of serotonin syndrome.

- **Over-the-counter supplements**: Certain supplements, like **St. John's Wort**, **tryptophan**, or **5-HTP (5-Hydroxytryptophan)**, can also increase serotonin levels.

- **Illegal drugs**: Substances such as **MDMA (ecstasy)**, **LSD**, and **cocaine** can cause serotonin levels to spike.

- **Certain pain medications**: Medications like **tramadol** and **fentanyl** have been associated with serotonin syndrome due to their serotonergic properties.

The risk of serotonin syndrome is particularly heightened when SSRIs are combined with other serotonergic drugs, either intentionally or unintentionally. Patients should always inform their healthcare providers about all medications, including over-the-counter drugs and supplements, they are currently taking to minimize the risk of serotonin syndrome.

Signs and Symptoms of Serotonin Syndrome

Serotonin syndrome can develop rapidly, often within hours of starting a new medication, increasing the dose, or adding another serotonergic drug. Symptoms can range from mild to severe and may include the following:

1. Neurological Symptoms

- **Agitation**: A feeling of restlessness or heightened anxiety.

- **Confusion**: Difficulty concentrating or thinking clearly.

- **Delirium**: Disorientation, hallucinations, or loss of touch with reality.

- **Tremors and muscle rigidity**: Involuntary shaking, jerking, or stiffness of the muscles.

- **Hyperreflexia**: Overactive reflexes, such as exaggerated or brisk responses to stimuli.

- **Clonus**: Rhythmic muscle contractions, often seen in the lower limbs.

2. Autonomic Symptoms

- **Hyperthermia**: Elevated body temperature, often exceeding 104°F (40°C), which can be dangerous and lead to organ damage.

- **Sweating**: Excessive perspiration, even without exertion.

- **Tachycardia**: Rapid heart rate (greater than 100 beats per minute).

- **Hypertension**: High blood pressure, which can be dangerous if left untreated.

- **Shivering**: Uncontrollable shaking or shivering.

3. Gastrointestinal Symptoms

- **Nausea and vomiting**: A common symptom in serotonin syndrome due to the disruption of normal serotonin levels.
- **Diarrhea**: Increased frequency of bowel movements due to the overstimulation of the gastrointestinal tract.

In severe cases, serotonin syndrome can lead to **seizures, rhabdomyolysis** (breakdown of muscle tissue), **renal failure**, and even death. The severity of the condition depends on the amount of serotonin excess in the brain and the speed at which the condition develops.

Diagnosing Serotonin Syndrome

Diagnosing serotonin syndrome can be challenging because its symptoms overlap with other conditions, such as infections, drug toxicity, and other medical emergencies. However, healthcare providers typically rely on a combination of a patient's medical history, physical examination, and symptom assessment to make a diagnosis.

A commonly used tool to assist in diagnosing serotonin syndrome is the **Hunter Serotonin Toxicity Criteria**. This diagnostic scale helps clinicians identify key signs and symptoms of serotonin toxicity, such as clonus, hyperreflexia, and fever. If these criteria are met and there is a history of serotonergic drug use or interactions, serotonin syndrome is likely.

Additionally, blood tests may be conducted to rule out other conditions that may mimic serotonin syndrome, such as infections or electrolyte imbalances.

How Serotonin Syndrome Is Treated and Prevented

The first step in treating serotonin syndrome is to immediately discontinue all serotonergic drugs, including SSRIs and any other medications or substances that may be contributing to the excess serotonin levels. Once the medication is stopped, treatment focuses on alleviating symptoms and stabilizing the patient.

Mild Cases

For mild cases of serotonin syndrome, patients may be monitored closely in a healthcare setting. In these cases, symptoms may improve within 24 to 48 hours once the offending drug is discontinued. Supportive care, such as cooling measures for hyperthermia and hydration, may be sufficient.

Moderate to Severe Cases

In moderate to severe cases, more intensive treatment may be necessary. This could involve:

- **Sedation**: Benzodiazepines (e.g., **lorazepam**) may be administered to reduce agitation, muscle rigidity, and hyperreflexia.

- **Serotonin antagonists**: In some cases, medications such as **cyproheptadine** (a serotonin antagonist) may be used to directly counteract the effects of excess serotonin.

- **Cooling measures**: In cases of hyperthermia, cooling blankets, ice packs, and other methods to lower body temperature are used.

- **Supportive care**: In some cases, intravenous fluids, electrolyte management, and even intubation or mechanical ventilation may be required if respiratory distress occurs.

Prevention

To prevent serotonin syndrome, it is essential to:

- **Avoid combining multiple serotonergic drugs**: Patients taking SSRIs should be cautious about combining them with other drugs that affect serotonin, such as MAOIs, SNRIs, or over-the-counter supplements.

- **Gradual dose adjustments**: When increasing the dose of an SSRI or introducing a new serotonergic medication, a gradual approach should be followed to minimize the risk of serotonin buildup.

- **Close monitoring**: Patients starting a new SSRI or adjusting their dosage should be closely monitored for any early signs of serotonin syndrome, particularly during the first few weeks of treatment.

- **Patient education**: Educating patients about the risks of combining serotonergic drugs and recognizing the symptoms of serotonin syndrome can help prevent and quickly identify the condition.

Conclusion

Serotonin syndrome is a potentially life-threatening condition that results from excessive serotonin activity in the brain. While it is rare, it is important for healthcare providers and patients to be aware of the risk, especially when SSRIs are used in combination with other serotonergic medications. Early recognition of symptoms and prompt discontinuation of the offending drugs are critical to managing serotonin syndrome successfully. By following appropriate precautions and educating patients about the risks and signs of serotonin syndrome, its occurrence can be minimized, allowing SSRIs to be used safely and effectively in the treatment of mental health disorders.

In the next chapter, we will explore how SSRIs interact with other medications, including potential drug interactions that may influence their effectiveness or increase the risk of adverse effects like serotonin syndrome. Understanding these interactions is key to optimizing treatment and minimizing risks for patients.

Chapter 10: Managing SSRIs with Other Medications

Selective serotonin reuptake inhibitors (SSRIs) are commonly prescribed to treat various mental health conditions, including depression, anxiety, obsessive-compulsive disorder (OCD), and panic disorders. However, as with any medication, SSRIs do not exist in isolation. Many patients with mental health conditions also have other physical or mental health concerns, which may require additional medications. This chapter focuses on the interactions between SSRIs and other medications, the importance of consulting with healthcare providers about all drugs being taken, and how SSRIs can be safely managed alongside other treatments.

Understanding Drug Interactions with SSRIs

SSRIs primarily work by inhibiting the reuptake of serotonin in the brain, increasing the amount of serotonin available for signaling between neurons. This modulation of serotonin levels can interact with other medications that influence serotonin, other neurotransmitters, or metabolic processes in the body. Some interactions can be beneficial, while others may be dangerous, leading to adverse effects, reduced efficacy, or even life-threatening conditions, such as serotonin syndrome.

Understanding these potential interactions is crucial for both healthcare providers and patients to ensure that SSRIs are used safely and effectively in combination with other medications. The following sections will highlight the most common and significant drug interactions with SSRIs, as well as strategies to mitigate risks.

Medications That Increase Serotonin Levels

1. Monoamine Oxidase Inhibitors (MAOIs)

MAOIs, such as **phenelzine** (Nardil) and **tranylcypromine** (Parnate), are older classes of antidepressants that also increase serotonin levels in the brain. When combined with SSRIs, there is a significant risk of **serotonin syndrome**, a potentially life-threatening condition caused by excessive serotonin levels.

Management:

- **Avoid co-prescription**: SSRIs should not be used concurrently with MAOIs. A washout period of at least 14 days is typically recommended between discontinuing an MAOI and starting an SSRI.
- **Close monitoring**: If an MAOI is being prescribed, it is essential that patients are carefully monitored for signs of serotonin syndrome, especially when transitioning medications.

2. Other SSRIs and Serotonergic Drugs

Using multiple serotonergic drugs simultaneously can increase serotonin levels to dangerous levels, leading to serotonin syndrome. These drugs include other SSRIs, serotonin-norepinephrine reuptake inhibitors (SNRIs), **tricyclic antidepressants (TCAs)**, and even certain **over-the-counter supplements** like **St. John's Wort** or **5-HTP**.

Management:

- **Medication review**: Always review all medications, including over-the-counter and herbal supplements, with a healthcare provider to avoid interactions that may increase serotonin.
- **Cautious prescribing**: If another serotonergic drug is necessary, alternatives with lower serotonin activity, such as **bupropion**, should be considered.

3. Triptans (for Migraines)

Triptans, like **sumatriptan** (Imitrex) and **rizatriptan** (Maxalt), are used to treat migraine headaches and also increase serotonin levels. Combining SSRIs with triptans increases the risk of serotonin syndrome, especially if high doses of both are used.

Management:

- **Cautious use**: If both medications are necessary, they should be used at the lowest possible effective doses and carefully monitored for any signs of serotonin syndrome.
- **Alternatives**: Non-serotonergic treatments for migraines, such as **propranolol** (a beta-blocker), may be considered as safer alternatives when possible.

Medications That Affect Cytochrome P450 Enzymes

SSRIs are metabolized in the liver by enzymes in the cytochrome P450 family, primarily **CYP2D6**, **CYP3A4**, and **CYP2C19**. Medications that affect these enzymes can either increase or decrease the levels of SSRIs in the body, potentially altering their efficacy or increasing the risk of side effects.

1. CYP2D6 Inhibitors

Some SSRIs, particularly **fluoxetine** (Prozac) and **paroxetine** (Paxil), are strong inhibitors of the **CYP2D6** enzyme. When taken in combination with other drugs metabolized by this enzyme, such as **tamoxifen** (used in breast cancer treatment) or certain antipsychotics like **risperidone**, the effectiveness of these drugs may be reduced.

Management:

- **Alternative medications**: For patients requiring tamoxifen, a non-CYP2D6-inhibiting SSRI, such as **sertraline** (Zoloft), is often preferred.
- **Dose adjustments**: In cases where a CYP2D6 inhibitor is necessary, dose adjustments for other medications metabolized by this enzyme may be required.

2. CYP3A4 Interactions

Other SSRIs, such as **sertraline** (Zoloft) and **escitalopram** (Lexapro), are less likely to inhibit CYP2D6 but can still interact with drugs metabolized by **CYP3A4**, including certain **antifungal medications** (e.g., **ketoconazole**), **antiviral drugs** (e.g., **ritonavir**), and **benzodiazepines** (e.g., **alprazolam**).

Management:

- **Monitoring levels**: For drugs that are metabolized by CYP3A4, such as alprazolam, healthcare providers may adjust dosages or switch to safer alternatives.
- **Close observation**: Patients on combinations that affect CYP3A4 should be carefully monitored for potential side effects or altered drug efficacy.

Medications That Decrease Serotonin Availability

Certain medications can decrease serotonin levels in the brain, potentially reducing the efficacy of SSRIs. These include **antipsychotic medications**, **stimulants**, and **corticosteroids**, which can sometimes block serotonin release or affect its availability in other ways.

1. Antipsychotic Medications

Some **atypical antipsychotics**, like **clozapine** (Clozaril), can interact with SSRIs, leading to reduced serotonin levels. The combination of these medications, particularly at higher doses, may also increase the risk of side effects such as sedation or weight gain.

Management:

- **Adjustment of dose**: In some cases, the dosage of the SSRI or the antipsychotic may need to be adjusted.
- **Close monitoring**: Regular monitoring for side effects, including weight gain, sedation, or metabolic changes, is advised when combining these types of medications.

2. Stimulants

Stimulant medications like **methylphenidate** (Ritalin) and **amphetamine salts** (Adderall), commonly prescribed for attention-deficit/hyperactivity disorder (ADHD), can reduce the effectiveness of SSRIs by interfering with serotonin function.

Management:

- **Careful monitoring**: When used together, it's important to monitor closely for symptoms of agitation, anxiety, or irritability, which may be exacerbated by the combination of stimulants and SSRIs.
- **Consider alternative treatments**: If ADHD treatment is needed, non-stimulant medications, such as **atomoxetine** (Strattera), may be considered.

Managing SSRIs with Medications for Other Conditions

In addition to the interactions discussed above, it's important to consider how SSRIs may interact with medications used for conditions such as **heart disease**, **diabetes**, **epilepsy**, and **gastrointestinal issues**. For example, SSRIs can sometimes interact with **blood thinners**, increasing the risk of bleeding, or with **anticonvulsants**, affecting seizure control.

1. Blood Thinners (Anticoagulants)

SSRIs, especially **paroxetine** (Paxil) and **fluoxetine** (Prozac), can interfere with platelet aggregation, potentially increasing the risk of bleeding when used in combination with anticoagulants like **warfarin**.

Management:

- **Monitoring**: Close monitoring of blood clotting times (INR) is necessary for individuals on both SSRIs and blood thinners.
- **Adjustment of dosage**: Adjusting the dosage of either the SSRI or anticoagulant may be necessary to ensure safety.

2. Antiepileptic Drugs

SSRIs can interact with antiepileptic drugs like **phenytoin** (Dilantin) and **carbamazepine** (Tegretol), affecting their metabolism. In some cases, this can lead to reduced efficacy or increased side effects.

Management:

- **Careful titration**: Both the SSRI and antiepileptic drug may require dosage adjustments.
- **Monitoring for side effects**: Patients should be monitored for changes in seizure frequency or other side effects.

The Importance of Consulting with a Healthcare Provider

Given the complexity of potential drug interactions, it is essential for patients to work closely with their healthcare providers when starting, stopping, or adjusting the use of SSRIs. Healthcare providers can help manage and adjust medications to avoid harmful interactions, optimize the efficacy of SSRIs, and ensure that patients receive the best possible treatment for their conditions.

Conclusion

Managing SSRIs alongside other medications requires careful consideration of potential drug interactions, both beneficial and harmful. Understanding the medications that increase or decrease serotonin levels, affect liver enzyme metabolism, or alter neurotransmitter function is key to ensuring the safe and effective use of SSRIs. By maintaining open communication with healthcare providers, patients can minimize risks and optimize the benefits of SSRIs in managing mental health conditions while addressing any other concurrent health issues. In the next chapter, we will explore how to safely begin SSRI treatment, including guidelines for dose adjustments and expectations during the early stages of therapy.

Chapter 11: How to Start Taking SSRIs

Starting treatment with selective serotonin reuptake inhibitors (SSRIs) can be a transformative step in managing mental health conditions, but it is important to approach it with the right knowledge and expectations. SSRIs, like all medications, require careful consideration to ensure that they are used effectively and safely. This chapter will provide guidelines for beginning SSRI treatment, the importance of gradual dosage increases, and what to expect during the early stages of treatment.

Guidelines for Beginning SSRI Treatment

When starting an SSRI, the process is not as simple as just taking the prescribed dose. Successful treatment begins with understanding the medication and its effects on both the body and the mind. Here are the essential steps for starting SSRI treatment:

1. Consultation with a Healthcare Provider

Before starting any SSRI, it is crucial to have a thorough discussion with a healthcare provider. This conversation should include:

- **Health history review**: Discussing any pre-existing medical conditions, including heart issues, liver or kidney problems, and other psychiatric conditions that could influence the choice of SSRI.
- **Medication history**: Informing the healthcare provider about any other medications being taken, including over-the-counter drugs and supplements, to prevent harmful interactions.
- **Lifestyle factors**: Understanding the patient's sleep habits, diet, and stress levels can also help tailor the treatment plan.

Your healthcare provider will help determine which SSRI is most appropriate based on your condition, overall health, and potential interactions with other medications. Once prescribed, the healthcare provider will give specific instructions on how to start the medication.

2. Understanding the Dosage

Typically, SSRIs are started at a **low dose** to minimize the risk of side effects. For example, for conditions like depression or anxiety, the initial dose might be lower than the recommended therapeutic dose. This gradual approach helps the body adjust to the medication and can prevent overwhelming side effects like nausea or dizziness.

- **Starting low, going slow**: The principle of "start low, go slow" refers to starting the medication at a low dose and gradually increasing it over time. This approach is used to minimize side effects and allow the brain to adapt to the increase in serotonin.

- **Regular follow-ups**: After starting treatment, it is important to follow up with your healthcare provider regularly. These check-ins allow them to assess how the medication is affecting you, whether any adjustments to the dosage are needed, and how you're responding emotionally and physically.

3. Initial Expectation Setting

Understanding what to expect at the start of SSRI treatment is crucial for patient compliance and mental well-being. Patients should be aware that while SSRIs can significantly improve mood and anxiety over time, the effects may not be immediate.

- **Delayed onset of therapeutic effects**: SSRIs generally take **2 to 4 weeks** to begin showing noticeable effects. For some patients, it may take up to **6 weeks** to feel the full benefits. It is important to stay patient and consistent with the prescribed medication regimen, even if immediate relief is not experienced.

- **Managing initial side effects**: The first few days or weeks of treatment are often marked by mild side effects such as nausea, headaches, or changes in sleep patterns. These symptoms are typically temporary and decrease as the body adjusts to the medication. If side effects persist or are intolerable, patients should consult their healthcare provider to assess whether the dosage should be adjusted or a different SSRI should be considered.

The Importance of Gradual Dosage Increases

A key aspect of starting SSRI treatment is the gradual increase in dosage. This is important for several reasons:

1. Minimizing Side Effects

When SSRIs are introduced into the body, the increase in serotonin can cause various side effects, such as dizziness, gastrointestinal discomfort, and insomnia. By starting with a low dose and gradually increasing it, patients give their body time to adjust, which can significantly reduce the severity of these side effects.

2. Enhancing Efficacy

For many SSRIs, the therapeutic effects become more pronounced with a gradual buildup of the medication in the body. Increasing the dose over time helps to achieve the optimal serotonin level in the brain, enhancing the medication's effectiveness.

3. Reducing Discontinuation Syndrome Risk

Some patients may experience symptoms when transitioning between different doses or when discontinuing SSRIs. A slow and gradual increase in dosage helps reduce the likelihood of these withdrawal symptoms, making the process of starting treatment smoother.

4. Personalized Adjustment

Each patient responds differently to SSRIs, and their ideal dosage may vary. Gradual dose escalation allows for personalized adjustments based on how the patient feels during treatment. This approach ensures that the medication is tailored to the patient's unique needs and response.

What to Expect in the Early Stages of Treatment

Starting an SSRI treatment regimen involves understanding both the potential benefits and the common experiences during the first few weeks. Here are key points to consider during the initial stages of SSRI treatment:

1. Onset of Side Effects

As mentioned earlier, some patients experience mild side effects during the first few days or weeks of SSRI use. Common side effects include:

- **Gastrointestinal issues**: Nausea, dry mouth, or mild stomach upset.
- **Sleep disturbances**: Either difficulty falling asleep or feeling overly drowsy.
- **Headaches or dizziness**: These side effects typically subside after a few days as the body adjusts.

These side effects are usually transient, but if they become severe or persistent, it is essential to communicate with the healthcare provider to adjust the dosage or try a different SSRI.

2. Emotional Adjustment Period

In some cases, patients may experience a brief emotional "bump" during the first few days of treatment, as the body and brain adjust to the increased serotonin levels. For some people, this adjustment period can lead to temporary mood swings, agitation, or heightened anxiety. This is usually a temporary phase that resolves as the medication takes effect.

Patience and self-care

3. Monitoring Progress

A key aspect of starting SSRI treatment is **monitoring progress**. The first few weeks are crucial for determining whether the SSRI is helping to alleviate symptoms like anxiety, sadness, or intrusive thoughts. Healthcare providers often schedule regular follow-up visits during this time to assess the patient's emotional state, side effects, and overall response to treatment.

- **Tracking symptoms**: Some individuals find it helpful to keep a journal or symptom diary to track changes in mood, sleep, appetite, and anxiety levels. This information can be valuable when discussing progress with a healthcare provider.
- **Evaluating effectiveness**: If there is no significant improvement in symptoms after 4 to 6 weeks, or if side effects remain intolerable, the provider may consider adjusting the dose, trying a different SSRI, or adding a complementary treatment.

Gradual Transition and Combination Therapy

Some patients may start with one SSRI but may require adjustments in the future. In certain cases, **combination therapy**, which involves using an SSRI in conjunction with other treatments such as **cognitive-behavioral therapy (CBT)** or other medications (like **benzodiazepines** for short-term anxiety relief), may be recommended. The process of adding or changing medications should always be done under the supervision of a healthcare provider.

Conclusion

Starting treatment with SSRIs is an essential step in the journey toward improved mental health, but it requires careful planning and patience. Beginning with a low dose, gradually increasing it, and managing side effects effectively are crucial to ensuring the best possible outcome. In the early stages, it is important for patients to stay informed, set realistic expectations, and maintain open communication with their healthcare provider. With proper management and patience, SSRIs can be an effective tool for addressing mental health issues, providing long-term benefits that significantly improve quality of life.

In the next chapter, we will delve into the timing and dosage considerations when using SSRIs. Understanding the recommended dosages and how to handle missed doses can play a crucial role in maximizing the medication's effectiveness and maintaining consistent progress in treatment.

Chapter 12: Timing and Dosage Considerations

One of the most critical aspects of SSRI treatment is adhering to the appropriate timing and dosage. SSRIs can be highly effective for managing conditions like depression, anxiety, and obsessive-compulsive disorder (OCD), but their success largely depends on taking them as prescribed, at the right times, and in the right amounts. In this chapter, we will explore the recommended dosages for various SSRIs, the importance of sticking to a prescribed schedule, and strategies for handling missed doses.

Recommended Dosages for Different SSRIs

While SSRIs generally share a common mechanism of action (increasing serotonin levels by inhibiting its reuptake), different SSRIs have unique dosing guidelines based on their pharmacokinetics and side effect profiles. The following are the typical starting doses for some of the most commonly prescribed SSRIs, though your healthcare provider may adjust these based on your specific needs, age, and response to the medication:

1. Fluoxetine (Prozac)

- **Starting dose**: 10–20 mg daily, typically taken in the morning.
- **Maintenance dose**: 20–40 mg per day (can be adjusted up to 80 mg daily, based on response).
- **Note**: Fluoxetine has a long half-life, meaning it stays in the system for an extended period and does not require as frequent dose adjustments. It is often preferred for patients who have issues with compliance, as it has a long-acting effect.

2. Sertraline (Zoloft)

- **Starting dose**: 25–50 mg daily, taken in the morning or evening.
- **Maintenance dose**: 50–200 mg per day.
- **Note**: Sertraline is commonly used for depression, anxiety disorders, and PTSD, with doses generally titrated upward over several weeks.

3. Paroxetine (Paxil)

- **Starting dose**: 10–20 mg daily, typically taken in the morning.
- **Maintenance dose**: 20–40 mg per day.
- **Note**: Paroxetine is often prescribed for anxiety disorders and depression. It is known to be more sedating than some other SSRIs, so it is frequently taken in the evening.

4. Citalopram (Celexa)

- **Starting dose**: 10–20 mg daily, taken in the morning.
- **Maintenance dose**: 20–40 mg per day (up to a maximum of 40 mg).
- **Note**: Citalopram is a widely prescribed SSRI for treating depression. It is generally considered one of the more tolerable SSRIs, with a relatively low risk of drug interactions.

5. Escitalopram (Lexapro)

- **Starting dose**: 10 mg daily, taken in the morning.
- **Maintenance dose**: 10–20 mg per day.
- **Note**: Escitalopram is the S-enantiomer of citalopram and is thought to be more potent with fewer side effects, making it a good option for many patients. It's commonly used for both depression and anxiety disorders.

The Importance of Sticking to the Prescribed Schedule

Consistency is key when it comes to SSRIs. Unlike some medications that work quickly, SSRIs require time to build up in the system and take effect. Inconsistent dosing can lead to fluctuations in serotonin levels, which can disrupt the medication's efficacy and may even trigger withdrawal-like symptoms, even if the patient misses a dose for just one day. Here are some key reasons why it's important to follow the prescribed dosing schedule:

1. Maintaining Steady Serotonin Levels

SSRIs work by increasing serotonin availability in the brain, but this process requires a consistent and steady presence of the medication in the system. Missing doses or taking them at irregular times can result in fluctuating serotonin levels, which may reduce the effectiveness of the drug and cause symptoms to return.

2. Minimizing Side Effects

Taking the medication at the same time each day helps the body get used to its effects. When doses are missed or taken erratically, side effects like nausea, headaches, and dizziness may occur more frequently or severely.

3. Enhancing Effectiveness

SSRIs generally take **2 to 4 weeks** to show noticeable effects, and consistent dosing is essential for achieving their full therapeutic potential. Skipping doses or delaying medication intake can delay the onset of benefits and lead to unnecessary frustration for patients hoping for quick relief.

The Timing of Doses

The timing of SSRI doses can be just as important as the dosage itself. While most SSRIs are taken once daily, there are a few key considerations related to timing that can affect both effectiveness and side effect management:

1. Morning vs. Evening Dosing

- **Morning**: Many people prefer to take SSRIs in the morning because some medications, like **fluoxetine** and **sertraline**, can have a stimulating effect that could interfere with sleep if taken too late in the day. Taking the medication in the morning helps individuals maintain their energy throughout the day.
- **Evening**: Some SSRIs, such as **paroxetine**, may cause drowsiness or sedation, making them better suited for evening dosing. If you experience sleep disturbances due to the medication, taking it before bed may help alleviate those effects.

2. With or Without Food

Most SSRIs can be taken with or without food, but some individuals find that taking the medication with food can help prevent gastrointestinal side effects like nausea or upset stomach. If you experience these issues, consider taking your SSRI with a meal.

How to Handle Missed Doses

Missing a dose of your SSRI can be concerning, but it's important to handle missed doses appropriately to minimize disruptions in treatment. Here's how to manage missed doses:

1. If a Dose is Missed by Less Than 12 Hours

If it has been less than **12 hours** since the scheduled dose, take the missed dose as soon as you remember. Do not double the dose to make up for the missed one. Instead, continue with your regular dosing schedule after taking the missed dose.

2. If a Dose is Missed by More Than 12 Hours

If more than **12 hours** have passed since the missed dose, skip the missed dose entirely and take the next dose at the regular time. Do not take two doses at once to compensate for the missed dose, as this can increase the risk of side effects.

3. Establishing a Routine

To reduce the likelihood of missing doses, try to establish a regular routine. Taking the medication at the same time each day—such as alongside breakfast or before bed—can help ensure that doses are not missed. Some people find it helpful to set a daily reminder on their phone or use a pillbox to organize their medications.

Adjusting Dosages During Treatment

During the course of SSRI treatment, your healthcare provider may adjust your dosage based on how you respond to the medication. Adjustments may be necessary if:

- The initial dose is not effective in controlling symptoms.

- Side effects become troublesome and need to be managed.

- There are changes in health conditions or medications that affect SSRI metabolism.

It's essential to follow your healthcare provider's guidance during this process and not to increase or decrease the dose on your own. Sudden changes in dosage can lead to side effects or reduce the medication's effectiveness.

Conclusion

Timing and dosage are fundamental to maximizing the benefits of SSRIs. By sticking to a consistent dosing schedule, adjusting doses gradually, and taking the medication at the optimal time for your needs, you can enhance the efficacy of your treatment and minimize potential side effects. Handling missed doses carefully, monitoring for side effects, and maintaining regular follow-ups with your healthcare provider will ensure that the medication works effectively for you.

In the next chapter, we will explore the timeline of SSRI effectiveness, including when you can expect to see the full benefits and how long treatment typically lasts. Additionally, we'll discuss the importance of continued evaluation and monitoring throughout your SSRI treatment.

Chapter 13: The Timeline of SSRI Effectiveness

One of the most common questions people have when starting an SSRI (Selective Serotonin Reuptake Inhibitor) is, "How long until I start feeling better?" While SSRIs are among the most effective treatments for depression, anxiety, and other mood disorders, they don't provide instant results. The process of SSRI effectiveness is gradual, and understanding the timeline can help patients manage their expectations and stay committed to the treatment. In this chapter, we will explore when to expect the full benefits of SSRIs, how long treatment typically lasts, and the importance of ongoing evaluation and monitoring to ensure the best outcome.

When to Expect the Full Benefits of SSRIs

SSRIs do not offer immediate relief for mental health conditions. Although many patients begin taking SSRIs with the hope of quickly feeling better, the effects usually take some time to become noticeable. Here is a general timeline for how long it typically takes for SSRIs to start working:

1. The First Few Days: Initial Adjustment

In the first few days of SSRI treatment, the body is adjusting to the medication, and most people do not experience significant improvements in mood or mental health. However, some early effects, such as changes in sleep patterns or minor side effects like nausea, can be noticed during this period. It is important to be aware that side effects in the initial stages, like fatigue, gastrointestinal issues, or dizziness, do not necessarily reflect the long-term benefits of the medication.

2. 1–2 Weeks: Early Signs of Improvement

After about 1 to 2 weeks of treatment, many people start to notice mild improvements in their symptoms. These early changes may include feeling slightly less anxious or more motivated. For some, the ability to focus or perform everyday tasks becomes easier. However, these early improvements are generally subtle, and it's important to continue taking the medication as prescribed, even if the changes are not dramatic.

3. 3–4 Weeks: Noticeable Improvement

By **3 to 4 weeks**, many individuals begin to experience more significant improvements in their mood, energy levels, and anxiety. For patients struggling with depression, they may start to feel a reduction in feelings of sadness or hopelessness. Those with anxiety disorders may find that their panic attacks or intrusive thoughts become less frequent and intense. At this point, it's essential to have a follow-up with the healthcare provider to discuss the progress and assess whether the dosage needs to be adjusted.

4. 6–8 Weeks: Full Therapeutic Effects

It often takes **6 to 8 weeks** for SSRIs to reach their full therapeutic effect. By this point, many patients experience substantial improvements in their symptoms. People with depression may begin to feel a renewed sense of well-being, and anxiety-related symptoms may decrease significantly. It's also around this time that patients may notice more profound changes in their overall emotional state, such as improved ability to manage stress, increased social engagement, and better sleep patterns.

However, it's important to note that not all patients respond at the same pace, and for some, the full benefit may take up to **12 weeks** or more. This variability depends on individual factors, including the specific SSRI being used, the severity of the condition, and the patient's unique neurochemistry.

How Long Treatment Typically Lasts

The length of time a person remains on an SSRI depends on several factors, including the nature of their mental health condition, how they respond to the medication, and whether they are experiencing any side effects. The treatment duration is typically broken down into three phases:

1. Acute Phase

The acute phase refers to the initial treatment period when SSRIs are introduced to stabilize mood and alleviate symptoms. This phase usually lasts for **6 to 12 weeks**, during which time the healthcare provider monitors the patient's response to the medication. Some individuals may begin to feel better in just a few weeks, while others may take longer.

2. Maintenance Phase

Once the acute symptoms have improved, the maintenance phase begins. During this phase, patients continue their SSRI treatment to maintain the benefits achieved in the acute phase and prevent the return of symptoms. The maintenance phase may last for **6 months to 1 year**, depending on the severity of the condition. For individuals with chronic depression or anxiety disorders, a longer maintenance phase may be necessary to keep symptoms under control.

3. Long-Term Use or Discontinuation

For some people, SSRI treatment may be needed for several years or even longer. This is especially true for those with recurrent depression or chronic anxiety disorders. Long-term treatment can help prevent relapses and manage ongoing symptoms. However, some individuals may successfully discontinue SSRIs after a period of stability, especially if they have worked with a therapist and developed coping strategies.

It's important to work closely with a healthcare provider when considering whether to continue or discontinue SSRIs. Discontinuing treatment should be done gradually under medical supervision to avoid withdrawal symptoms or a relapse of the original condition.

The Importance of Continued Evaluation and Monitoring

Throughout SSRI treatment, regular evaluation and monitoring are critical to ensure that the medication is effective and that any side effects are managed appropriately. Here's why monitoring is so important:

1. Assessing Medication Effectiveness

In the early stages of treatment, healthcare providers often schedule follow-up appointments to monitor progress. These visits are essential for:

- Evaluating symptom improvement: Is the medication improving mood, energy, and anxiety symptoms as expected?
- Adjusting dosage: If there has been insufficient improvement, the healthcare provider may consider increasing the dose or trying a different SSRI.

2. Managing Side Effects

Many patients experience side effects during the first few weeks of taking SSRIs. Regular monitoring helps identify side effects early so they can be addressed:

- Adjusting timing or dosage: If side effects are intolerable, the healthcare provider may adjust the timing of the medication or switch to a different SSRI.
- Adding adjunct treatments: For side effects like insomnia, additional treatments or behavioral strategies can be recommended.

3. Long–Term Considerations

For patients on SSRIs for extended periods, ongoing evaluation is essential to:

- Prevent relapse: Continuing to monitor for any signs of symptom recurrence ensures that any necessary adjustments to the treatment plan are made promptly.
- Addressing any changes in health: As patients age or experience changes in their health, their response to SSRIs may shift. Regular check-ins allow for adjustments to account for these changes.

Managing Expectations and Patience During Treatment

It's important for patients to maintain realistic expectations during SSRI treatment. The road to recovery is often gradual, and the timeline for SSRI effectiveness varies from person to person. Patience is a crucial component in the treatment process. For those who experience mild improvements within a few weeks, it's important to remember that full therapeutic benefits may still be several weeks away.

If you find that symptoms do not improve or side effects become overwhelming, don't hesitate to contact your healthcare provider. They can make adjustments to your treatment plan, whether it's increasing the dose, switching medications, or adding therapy or lifestyle changes.

Conclusion

Understanding the timeline of SSRI effectiveness is vital for managing expectations and ensuring successful treatment. While the therapeutic effects of SSRIs may take time to fully manifest, the gradual improvement that many individuals experience can significantly enhance their quality of life. Regular monitoring, adjusting dosages, and being patient with the process can help patients achieve the best possible outcomes. In the next chapter, we will explore the complementary role of **therapy** in SSRI treatment, and how combining SSRIs with cognitive-behavioral therapy (CBT) or other forms of psychotherapy can enhance recovery and overall mental health.

Chapter 14: The Role of Therapy in Combination with SSRIs

While SSRIs (selective serotonin reuptake inhibitors) have become a cornerstone of treatment for a wide range of mental health conditions, their effectiveness is often enhanced when combined with psychotherapy. Cognitive-behavioral therapy (CBT), in particular, has proven to be a highly effective complement to SSRI treatment, helping individuals not only manage their symptoms but also develop long-term strategies for mental health and emotional regulation. This chapter will explore how therapy works in tandem with SSRIs, the synergistic effects of combining medication and therapy, and best practices for creating a combined treatment plan for optimal mental health recovery.

How Cognitive Behavioral Therapy (CBT) Complements SSRI Treatment

Cognitive-behavioral therapy is a structured, goal-oriented form of psychotherapy that focuses on identifying and changing unhelpful thought patterns, behaviors, and emotional responses. CBT is particularly effective for treating conditions like depression, anxiety disorders, OCD, and PTSD—conditions commonly treated with SSRIs. When used in combination with SSRIs, CBT can provide patients with the tools to better understand and manage their thoughts and behaviors, further enhancing the benefits of medication.

1. Addressing Negative Thought Patterns

SSRI treatment helps increase serotonin levels in the brain, which can improve mood and reduce anxiety, but this alone may not address the deep-seated cognitive patterns that contribute to mental health disorders. Negative thought patterns, such as catastrophizing (expecting the worst outcome), black-and-white thinking (seeing situations as all good or all bad), and rumination (repetitively thinking about distressing thoughts) are common in conditions like depression and anxiety.

CBT works by teaching patients how to recognize these patterns and challenge them. For example, a patient might learn to reframe negative thoughts, replacing "I will never get better" with "I may struggle at times, but I can improve with effort and support." This process of cognitive restructuring complements the mood-enhancing effects of SSRIs by addressing the underlying cognitive distortions that contribute to mental health struggles.

2. Developing Coping Skills

In addition to modifying negative thinking, CBT equips individuals with practical coping strategies to manage stress, anxiety, and other difficult emotions. Techniques such as mindfulness, relaxation exercises, and grounding exercises can reduce the physical and emotional impact of anxiety or panic. These coping strategies not only help reduce the immediate distress caused by symptoms but also build emotional resilience for the long term.

Combining SSRIs with the development of coping mechanisms through CBT creates a more well-rounded approach to mental health. While SSRIs address the biochemical imbalances that contribute to conditions like depression and anxiety, CBT offers patients the tools to respond more effectively to life's challenges, leading to more sustainable improvements in mental health.

3. Enhancing Long–Term Mental Health

One of the biggest advantages of combining SSRIs with CBT is the potential for long-term mental health benefits. While SSRIs can alleviate symptoms quickly, they don't teach lasting coping strategies or provide the skills necessary to prevent relapse once medication is discontinued. Therapy, on the other hand, helps individuals understand the underlying causes of their mental health conditions and empowers them with the tools to manage their symptoms independently.

For example, individuals who complete a course of CBT while on SSRIs may experience a lower risk of relapse after discontinuing the medication. They will have learned strategies to prevent negative thought patterns and to handle difficult emotions on their own, which can be invaluable for maintaining mental well-being once the medication is tapered down or stopped altogether.

The Synergy of Medication and Therapy in Mental Health Recovery

The combination of SSRIs and therapy offers several advantages over either treatment alone, leading to faster and more comprehensive recovery. While SSRIs work by stabilizing serotonin levels and helping improve mood and emotional regulation, therapy helps individuals gain insight into the causes of their conditions and develop strategies to improve coping and emotional resilience.

1. Faster Symptom Relief

While SSRIs may take **2 to 4 weeks** to start showing noticeable benefits, therapy can begin to offer improvements more quickly. Therapy provides patients with tools and strategies to begin addressing their symptoms immediately, helping them feel more in control while they wait for the SSRI to reach its full effectiveness. This can provide a sense of empowerment and hope during the early stages of treatment.

2. Comprehensive Emotional Support

SSRIs can help regulate mood and reduce symptoms of anxiety and depression, but they do not necessarily address the emotional and behavioral aspects of these conditions. Therapy, particularly CBT, focuses on these aspects, helping individuals explore and modify their behaviors and reactions to life events. This dual approach ensures that patients receive both biological and emotional support, leading to a more holistic and sustainable recovery.

3. Prevention of Relapse

One of the key benefits of combining SSRIs with therapy is the potential for preventing relapse. SSRIs can be highly effective for managing symptoms in the short term, but without therapy, patients may not fully address the cognitive and behavioral patterns that contributed to their mental health struggles in the first place. By combining SSRIs with therapy, individuals can tackle both the symptoms and the root causes of their conditions, significantly reducing the chances of relapse once they discontinue treatment.

Best Practices for Combined Treatment Plans

To maximize the benefits of both SSRIs and therapy, it's important to follow best practices for combined treatment. These practices ensure that both the medication and therapy work synergistically and provide the best possible outcomes for mental health recovery.

1. Open Communication Between Healthcare Providers

Patients should ensure that their prescribing healthcare provider (such as a psychiatrist or primary care doctor) and their therapist (such as a licensed psychologist or licensed clinical social worker) are communicating regularly. This collaboration helps to align the goals of both treatment methods and ensures that the patient's progress is being monitored from both the medical and therapeutic perspectives. This may include discussions about the patient's progress, side effects of medication, and the development of coping strategies in therapy.

2. Consistent Engagement in Therapy

For therapy to be effective, it requires consistent engagement. While SSRIs may begin showing results within a few weeks, therapy benefits accumulate over time, with lasting results achieved through continued effort. Patients should commit to attending therapy sessions regularly, completing assigned homework (such as journaling or practicing new coping strategies), and actively participating in the process of self-discovery and personal growth.

3. Tailoring the Treatment Plan to the Individual

Each person's experience with mental health and recovery is unique, so treatment plans should be tailored accordingly. The combination of SSRIs and therapy can be adapted to address individual needs, such as incorporating specific therapeutic modalities (e.g., **exposure therapy** for PTSD or **dialectical behavior therapy** for borderline personality disorder) or adjusting SSRI dosages based on side effects and symptom relief.

4. Gradual Approach to Medication and Therapy

For patients who are new to treatment, it may be best to gradually introduce both SSRIs and therapy. Starting therapy early, while beginning the medication, can help patients build coping skills that enhance the effects of the SSRI. Additionally, if therapy alone is proving insufficient, SSRIs may be added after a period of therapy to further stabilize mood and reduce anxiety.

Conclusion

Combining SSRIs with therapy, particularly cognitive-behavioral therapy (CBT), offers a comprehensive and effective approach to mental health recovery. While SSRIs address the biochemical underpinnings of conditions like depression and anxiety, therapy provides the tools for individuals to understand, manage, and cope with their emotions and behaviors. The synergy between medication and therapy leads to faster symptom relief, long-term benefits, and a significantly lower risk of relapse. By working closely with both their healthcare provider and therapist, patients can optimize their treatment, ensuring that they receive the best possible care for their mental health and wellness.

In the next chapter, we will explore the role of SSRIs in special populations, including children, older adults, and individuals who are pregnant or breastfeeding. This chapter will provide valuable insights into how SSRI treatment can be adapted to meet the needs of these specific groups while maintaining safety and effectiveness.

Chapter 15: SSRIs and Special Populations

While SSRIs (Selective Serotonin Reuptake Inhibitors) are effective for a broad range of mental health conditions, special care must be taken when considering their use in specific populations. Children, adolescents, older adults, and pregnant or breastfeeding individuals may have different responses to SSRIs, both in terms of efficacy and safety. Understanding these differences and approaching SSRI treatment with tailored strategies is essential to ensuring the safety and success of therapy. This chapter explores the considerations and best practices for using SSRIs in these special populations.

1. SSRIs in Children and Adolescents

The use of SSRIs in children and adolescents has been a topic of much debate. Depression, anxiety, and obsessive-compulsive disorder (OCD) are increasingly diagnosed in younger populations, and SSRIs are commonly prescribed as part of the treatment plan. However, special considerations must be made when treating younger individuals with SSRIs.

a. Efficacy in Children and Adolescents

SSRIs have been shown to be effective for treating depression and anxiety in children and adolescents, although the response may not always be as rapid or pronounced as in adults. For instance, fluoxetine (Prozac) is the most commonly prescribed SSRI for younger patients, and studies have supported its use in treating major depressive disorder and obsessive-compulsive disorder in children over the age of 8.

However, SSRIs can take several weeks to show their full effect, and improvements in mood may not be immediately noticeable. Parents and caregivers should be prepared for a gradual response and maintain realistic expectations.

b. Risks and Side Effects

One of the major concerns when prescribing SSRIs to children and adolescents is the increased risk of suicidal thoughts or behaviors, especially during the early stages of treatment. This risk is more prominent in patients under 18, and careful monitoring is necessary. While SSRIs are generally safe, mental health professionals will closely assess the potential risks and benefits, particularly in the first few weeks of treatment.

Side effects in children and adolescents may also differ from those in adults. Common side effects include sleep disturbances, appetite changes, gastrointestinal issues, and agitation. If side effects become overwhelming, adjusting the dosage or switching medications may be necessary.

c. Best Practices for Treatment

- **Close Monitoring**: Children and adolescents on SSRIs should be closely monitored for changes in mood, behavior, and any signs of suicidal thoughts or self-harm.

- **Psychotherapy**: Combining SSRIs with psychotherapy, particularly cognitive-behavioral therapy (CBT), is highly beneficial for younger patients, helping them build coping skills and emotional resilience.

- **Parental Involvement**: Parents should be actively involved in the treatment plan, including attending appointments and helping with monitoring symptoms at home.

2. SSRIs in Older Adults

Older adults are one of the fastest-growing groups prescribed SSRIs, as they are commonly used to treat depression, anxiety, and other mood disorders. However, prescribing SSRIs for elderly patients requires special attention due to the potential for increased side effects and interactions with other medications.

a. Changes in Pharmacokinetics

As individuals age, their bodies experience changes in drug metabolism and clearance. The liver and kidneys, which play a role in metabolizing and excreting medications, may not function as efficiently, leading to higher blood concentrations of SSRIs and an increased risk of side effects. SSRIs such as **paroxetine** and **fluoxetine** are metabolized more slowly in older adults, requiring lower initial doses and more careful monitoring.

b. Potential Drug Interactions

Older adults often take multiple medications for various health conditions, increasing the potential for drug interactions. SSRIs can interact with medications such as blood thinners, antihypertensives, and other psychotropic drugs. For example, SSRIs like **sertraline** or **fluoxetine** may interact with anticoagulants like warfarin, increasing the risk of bleeding.

Healthcare providers need to carefully evaluate all medications the patient is taking and may adjust dosages or select a different SSRI to minimize risks.

c. Risks of Side Effects

Older adults may experience side effects such as dizziness, falls, or confusion. Some SSRIs, especially those with sedative effects like **paroxetine**, may increase the risk of these symptoms. Additionally, there may be an increased risk of **hyponatremia** (low sodium levels) in older adults taking SSRIs, which can lead to confusion, seizures, or even coma.

d. Best Practices for Treatment

- **Start with a Low Dose**: Due to changes in drug metabolism, starting with a lower dose and gradually increasing it can help minimize side effects.
- **Monitor for Side Effects**: Regular follow-up appointments are essential to monitor for side effects, such as dizziness, confusion, or weight changes.
- **Review All Medications**: A thorough review of the patient's entire medication regimen is essential to avoid drug interactions and ensure the SSRIs are not interfering with other treatments.

3. SSRIs During Pregnancy and Breastfeeding

The decision to use SSRIs during pregnancy and breastfeeding involves balancing the benefits of treating maternal mental health with the potential risks to the developing baby or infant. This decision should be made collaboratively by the patient and their healthcare provider, with careful consideration of the risks and benefits.

a. SSRIs and Pregnancy

During pregnancy, untreated depression or anxiety can significantly impact both the mother and the fetus, potentially leading to complications such as preterm birth, low birth weight, and poor maternal health. However, some SSRIs are considered safer during pregnancy than others.

- **First Trimester**: The first trimester is the most critical time for fetal development, and the use of SSRIs during this period has been linked to a small increased risk of birth defects, especially heart defects (e.g., with **paroxetine**). However, newer studies show that the overall risk of major malformations is low.
- **Third Trimester**: Some SSRIs, especially **fluoxetine**, have been linked to an increased risk of neonatal withdrawal symptoms, including irritability, feeding difficulties, and respiratory distress. These effects are typically mild and resolve over time but should be closely monitored.

b. SSRIs and Breastfeeding

SSRIs can pass into breast milk, but most are considered safe for breastfeeding mothers. **Sertraline** and **paroxetine** are often preferred as they have lower levels in breast milk and fewer potential effects on the infant. However, any side effects experienced by the infant—such as sedation, feeding difficulties, or irritability—should be promptly discussed with a healthcare provider.

c. Best Practices for Treatment

- **Close Monitoring**: During pregnancy, especially in the first and third trimesters, monitoring for any potential side effects is essential. If there are concerns about birth defects or neonatal withdrawal, the healthcare provider may recommend changing medications or reducing the dosage.
- **Psychotherapy**: In some cases, psychotherapy may be used as an adjunct or alternative to medication, especially if the risks of SSRI use during pregnancy are deemed too high.

4. Conclusion

The use of SSRIs in special populations requires careful consideration of the individual's unique circumstances, including age, pregnancy, and the presence of other health conditions. Whether it's for children, older adults, or pregnant and breastfeeding individuals, SSRIs can provide effective treatment for mental health conditions when managed appropriately. Collaboration between the patient, their family (if applicable), and healthcare providers is key to ensuring that SSRIs are used safely and effectively in these populations.

In the next chapter, we will explore the process of safely discontinuing SSRIs, including how to manage withdrawal symptoms and the importance of tapering off medication gradually.

Chapter 16: Understanding the Discontinuation Process

One of the most critical aspects of SSRI treatment is knowing how and when to safely discontinue use. While SSRIs are effective in managing conditions like depression, anxiety, and OCD, there comes a time when patients may no longer need to be on these medications, or they may need to reduce their dosage for other reasons. Discontinuing SSRIs must be done carefully and gradually, as abrupt cessation can lead to withdrawal symptoms, a risk that must be carefully managed to avoid setbacks in treatment or the recurrence of symptoms. This chapter will outline the discontinuation process, potential withdrawal symptoms, and the importance of tapering off SSRIs under professional guidance.

1. The Importance of Tapering Off SSRIs

Tapering refers to the gradual reduction of a medication dose over a period of time, allowing the body to adjust to the lower levels of the drug. When it comes to SSRIs, discontinuation should not be abrupt. Doing so can lead to what is commonly referred to as **SSRI discontinuation syndrome** or withdrawal, which can include a variety of physical and emotional symptoms.

Why Tapering is Necessary

SSRIs increase the availability of serotonin in the brain by inhibiting its reuptake. Over time, the brain adapts to these increased levels. Sudden cessation of SSRIs can disrupt the balance that the brain has adjusted to, leading to a rebound of symptoms such as anxiety or depression. Tapering allows the brain to slowly readjust, minimizing withdrawal symptoms and reducing the chances of a relapse.

2. Withdrawal Symptoms

Withdrawal symptoms from SSRIs are common, especially if the medication is stopped suddenly or reduced too quickly. These symptoms can vary in severity depending on the type of SSRI, the dosage, and how long the person has been taking the medication. Understanding these potential symptoms can help patients be prepared and discuss any concerns with their healthcare provider before discontinuation.

Common Withdrawal Symptoms

- **Flu-like symptoms**: This includes nausea, headaches, dizziness, muscle aches, and fatigue, which may mimic the symptoms of a viral infection.

- **Gastrointestinal disturbances**: Patients may experience symptoms like diarrhea or upset stomach, which can make the discontinuation process uncomfortable.

- **Insomnia and sleep disturbances**: Difficulty sleeping or experiencing vivid dreams is common when reducing or stopping SSRIs.

- **Mood swings**: Irritability, sadness, or even feelings of extreme anxiety may emerge during withdrawal. In some cases, patients may feel a resurgence of depression.

- **Electric shock sensations**: Often referred to as "brain zaps," these sensations are a distinct and common symptom during SSRI withdrawal. It feels like a sudden jolt or shock running through the head, which can be disorienting and uncomfortable.

- **Dizziness or lightheadedness**: Some patients report feeling dizzy or unsteady during SSRI tapering, especially when standing up.

- **Cognitive effects**: People may also experience difficulty concentrating, confusion, or "brain fog."

3. The Tapering Process: How It Works

Discontinuing SSRIs should always be done under the guidance of a healthcare provider. The process typically involves slowly reducing the dosage over weeks or even months, depending on how the individual is responding to the medication. Here's how it typically works:

Step 1: Gradual Dose Reduction

The first step in tapering off SSRIs is to reduce the dose slowly. The healthcare provider will typically recommend a gradual reduction of the daily dose by **10-25%** each week, depending on the medication and the patient's individual response. For example, if a patient is taking 40 mg of fluoxetine (Prozac) daily, the dose might be reduced to 30 mg for a week or two, then further decreased to 20 mg, and so on.

Step 2: Monitoring for Symptoms

During the tapering process, it's important for the patient to regularly check in with their healthcare provider to monitor any withdrawal symptoms or the return of the original condition (such as depression or anxiety). If symptoms are significant or the patient is experiencing considerable discomfort, the tapering process may be slowed down, or the dose may be increased temporarily to alleviate withdrawal.

Step 3: Transitioning to Lower Doses

For some SSRIs, such as fluoxetine, the medication is available in liquid form, allowing for even more precise dose adjustments as part of the tapering process. Some individuals may require extended time at lower doses before discontinuing the medication completely.

Step 4: Final Stages of Discontinuation

Once the dose is sufficiently reduced, and the patient has been symptom-free at lower doses for a while, the final step is discontinuing the SSRI completely. At this point, a healthcare provider may recommend tapering every few days or even stopping medication entirely, depending on the patient's progress.

4. Managing Withdrawal Symptoms

The symptoms associated with SSRI withdrawal can be managed in several ways. Here are some strategies for handling the most common symptoms:

1. Gradual Tapering

The best way to manage withdrawal symptoms is by tapering off the medication gradually. Slowing down the reduction process will help the body adjust to the lower serotonin levels and reduce the severity of withdrawal symptoms.

2. Supportive Care

If flu-like symptoms, nausea, or headaches occur, over-the-counter remedies or prescription medications for pain management and nausea can be used as recommended by a healthcare provider. Staying hydrated, eating nutritious meals, and getting plenty of rest can also help the body cope with withdrawal.

3. Mindfulness and Relaxation Techniques

For symptoms like anxiety, irritability, or mood swings, practicing mindfulness meditation, relaxation exercises, or deep breathing can help ease discomfort. Engaging in regular physical activity or yoga can also reduce stress and improve overall emotional regulation during the tapering process.

4. Psychological Support

Withdrawal can also affect a person's mental health. Talking to a therapist or counselor during this time can provide emotional support and help with any resurfacing feelings of anxiety or depression. Cognitive-behavioral therapy (CBT) or other forms of psychotherapy can also be helpful in providing coping strategies during the discontinuation process.

5. When to Seek Help

If withdrawal symptoms become severe or if the original condition (e.g., depression or anxiety) re-emerges, it's important to contact a healthcare provider immediately. In some cases, the tapering process may need to be adjusted, or additional interventions may be required. This may involve extending the tapering period or temporarily resuming the medication until symptoms are under control.

6. The Role of Therapy During Discontinuation

Even after discontinuing SSRIs, therapy can play a significant role in maintaining mental health stability. Cognitive-behavioral therapy (CBT), mindfulness-based therapy, or other approaches can help individuals learn new coping strategies, process emotions, and reduce the risk of relapse.

Continued therapy during and after SSRI discontinuation provides tools for maintaining mental well-being without relying on medication. This holistic approach can be empowering for patients who have successfully weaned off SSRIs.

7. Conclusion

Discontinuing SSRIs is a process that requires patience, planning, and the support of healthcare professionals. By tapering the medication gradually and managing withdrawal symptoms effectively, patients can safely transition off SSRIs while minimizing the risk of relapse or withdrawal. It is essential to understand that discontinuation is a highly individual process, and what works for one person may not work for another. Ongoing support from medical professionals, including monitoring for any signs of relapse or new symptoms, can help ensure a smooth transition and long-term mental health success. In the next chapter, we will explore strategies for overcoming treatment resistance in those who may not respond well to SSRIs.

Chapter 17: Overcoming Treatment Resistance with SSRIs

While SSRIs (Selective Serotonin Reuptake Inhibitors) are often highly effective in managing mental health conditions such as depression and anxiety, not all individuals respond to these medications in the way they should. In some cases, individuals may experience little or no improvement in their symptoms despite taking SSRIs as prescribed. This phenomenon, known as **treatment resistance**, is a significant challenge for both patients and healthcare providers. However, understanding the reasons behind treatment resistance and knowing the alternatives can help individuals find a path to recovery.

This chapter will explore why some people may not respond to SSRIs, strategies to overcome treatment resistance, and the role of alternative treatments in cases where SSRIs are ineffective.

1. What is Treatment Resistance?

Treatment resistance occurs when a patient's symptoms do not improve significantly after adequate treatment with one or more medications. In the context of SSRIs, this means that even after following the prescribed dosage for a reasonable amount of time, the patient does not experience the expected benefits in mood or anxiety reduction.

For depression, the standard rule is that an individual should be on an SSRI for at least **4-6 weeks** to assess its effectiveness. However, if after this time, there is no improvement or even a worsening of symptoms, it is considered treatment-resistant depression (TRD). Similarly, resistance can occur in individuals with anxiety disorders, obsessive-compulsive disorder (OCD), or post-traumatic stress disorder (PTSD) who do not respond to SSRIs after a sufficient trial period.

2. Why Do Some People Experience Treatment Resistance?

There are several reasons why some people may not respond to SSRIs. These reasons can be biological, genetic, or related to how the medication is processed by the body. Some of the common factors contributing to treatment resistance include:

a. Genetic Factors

One of the most important factors in determining how well a person will respond to SSRIs is their **genetic makeup**. Some individuals may have genetic variations that affect how their body metabolizes the drug. For example, variations in genes like **CYP450**, which plays a role in drug metabolism, can lead to SSRIs being metabolized too quickly or too slowly. This can either reduce their effectiveness or lead to side effects that prevent the medication from being tolerated.

b. Neurobiology of Depression and Anxiety

Mental health conditions like depression and anxiety are complex and involve multiple areas of the brain, neurotransmitter systems, and signaling pathways. While SSRIs primarily target the serotonin system, they may not address other contributing factors, such as imbalances in other neurotransmitters like dopamine, norepinephrine, or GABA (gamma-aminobutyric acid). If these pathways are also disrupted, SSRIs alone may not be sufficient to relieve symptoms.

c. Co-occurring Medical Conditions

Some individuals have other medical conditions that affect their response to SSRIs. For example, thyroid imbalances, chronic pain, or vitamin D deficiency can exacerbate depression symptoms or interfere with the effectiveness of antidepressants. Additionally, conditions like **bipolar disorder**, where mood swings occur between depression and mania, may require medications other than SSRIs.

d. Medication Interactions

If an individual is taking multiple medications, there could be **drug interactions** that impact the effectiveness of SSRIs. For example, certain medications used to treat pain, heart disease, or other psychiatric conditions can interfere with how SSRIs work. If these interactions are not recognized, treatment may not be optimized.

3. Strategies to Overcome Treatment Resistance

If SSRIs do not work, there are several strategies that healthcare providers can consider to overcome treatment resistance. Each individual's response to treatment will be different, and sometimes finding the right approach may require a bit of trial and error. Below are common strategies for overcoming SSRI treatment resistance:

a. Dose Adjustment

One of the first strategies to try is **increasing the dose** of the SSRI. Some patients who do not respond to the initial dose may benefit from a gradual increase in the dosage, which allows for better symptom control. However, it is crucial that dose increases are done under the supervision of a healthcare provider to avoid side effects.

b. Switching to a Different SSRI or Antidepressant

If increasing the dose does not work, another approach is to switch to a different SSRI. Although SSRIs generally have a similar mechanism of action, there are variations in how different SSRIs interact with serotonin receptors, and some individuals may respond better to one specific SSRI over another. For example, **escitalopram** may be more effective for some patients than **fluoxetine**, due to differences in receptor binding affinities and side effect profiles.

If switching between SSRIs doesn't yield results, it may be necessary to try **other classes of antidepressants**, such as **SNRIs** (serotonin-norepinephrine reuptake inhibitors), **tricyclic antidepressants** (TCAs), or **MAOIs** (monoamine oxidase inhibitors), which work through different mechanisms.

c. Augmenting Treatment with Other Medications

Augmentation involves adding another medication to the existing SSRI regimen. This can help target different neurotransmitter systems and improve treatment response. Common augmentation strategies include:

- **Adding an atypical antipsychotic**: Medications like **aripiprazole** or **quetiapine** can be added to SSRIs for individuals who don't respond to antidepressants alone. These medications can help with mood stabilization and further alleviate depressive or anxiety symptoms.
- **Adding a mood stabilizer**: For individuals with mood fluctuations or who may have underlying bipolar disorder, adding a **lithium** or **valproate** may be beneficial.
- **Adding a stimulant**: In some cases, especially for patients who also experience attention-deficit hyperactivity disorder (ADHD), adding a stimulant like **methylphenidate** may improve focus and energy levels.

d. Cognitive Behavioral Therapy (CBT)

While medications can be incredibly effective for treating depression and anxiety, **therapy** often plays an important role in overcoming treatment resistance. **Cognitive Behavioral Therapy (CBT)**, for example, can help individuals identify and change unhelpful thought patterns, develop coping strategies, and address the underlying psychological issues that may be contributing to their symptoms.

Combining SSRIs with therapy often leads to better outcomes, as the combination of medication and psychological support can address both the biological and emotional aspects of mental health disorders.

e. Psychotherapy or Alternative Therapies

In cases of severe treatment resistance, exploring alternative therapies like **psychedelic-assisted therapy** or **electroconvulsive therapy (ECT)** may be an option. While these therapies are not universally effective, they have shown promise in certain cases of treatment-resistant depression.

4. Alternative Treatments for SSRI–Resistant Cases

In addition to the augmentation strategies discussed above, there are other treatments to consider when SSRIs fail to provide relief:

a. Ketamine and Esketamine

Ketamine, once used primarily as an anesthetic, has gained attention for its rapid-acting antidepressant effects. **Esketamine**, a nasal spray form of ketamine, has been approved for use in treatment-resistant depression. Ketamine works differently from SSRIs by modulating **glutamate** levels in the brain, which can have a rapid effect on mood.

b. Transcranial Magnetic Stimulation (TMS)

Transcranial Magnetic Stimulation (TMS) is a non-invasive procedure that uses magnetic fields to stimulate specific areas of the brain involved in mood regulation. TMS is often used in patients who have not responded to traditional treatments like SSRIs. It has shown positive results in many individuals with treatment-resistant depression.

5. Conclusion

Treatment resistance is a frustrating challenge in the management of mental health conditions, but it is important to remember that SSRIs are not the only option. Through careful monitoring and the use of alternative therapies and medications, individuals can find an effective treatment path. Working closely with healthcare providers, exploring different approaches, and understanding that overcoming treatment resistance may take time are key to successfully managing mental health disorders.

In the next chapter, we will delve into how lifestyle factors such as diet, exercise, and sleep can significantly impact the effectiveness of SSRIs and contribute to overall mental wellness.

Chapter 18: The Future of SSRIs and Mental Health Treatment

The field of mental health treatment is constantly evolving, and SSRIs (Selective Serotonin Reuptake Inhibitors) have been at the forefront of this transformation. However, as our understanding of the brain and mental health disorders continues to grow, the landscape of treatment options is also changing. This chapter will explore the future of SSRIs, emerging trends in antidepressant research, and what innovations in neuroscience and pharmacology might mean for the next generation of mental health treatments.

1. New Frontiers in SSRI Research

Although SSRIs have been a cornerstone in treating depression and anxiety since the 1980s, researchers continue to explore new ways to enhance their efficacy and minimize their side effects. Recent advances in neurobiology and psychopharmacology have revealed new targets for antidepressant drugs, and SSRIs themselves may evolve in terms of their formulations and mechanisms of action.

Personalized Medicine

One promising area of research is the development of **personalized or precision medicine**. By using genetic testing and biomarkers, healthcare providers may be able to better match SSRIs (or alternative treatments) to individual patients based on their genetic makeup, improving the likelihood of success and minimizing side effects. This approach has the potential to revolutionize how we think about depression treatment, offering more tailored solutions for patients.

Novel SSRI Derivatives

Researchers are investigating modified versions of SSRIs that can target serotonin more specifically or act on multiple neurotransmitter systems simultaneously. These new drugs aim to improve the effectiveness of SSRIs for treatment-resistant patients or reduce side effects associated with traditional SSRIs, such as sexual dysfunction or weight gain.

2. Exploring Alternatives to SSRIs

As effective as SSRIs have been for many, they are not the panacea for everyone. The search for alternative treatments is ongoing, and many potential therapies show promise. Here are some emerging options:

Ketamine and Esketamine

Ketamine, a dissociative anesthetic, has emerged as a fast-acting treatment for severe depression, especially in cases that have not responded to other medications. Esketamine, a nasal spray form of ketamine, has been FDA-approved for treatment-resistant depression. While ketamine is not a serotonin-based treatment, it provides rapid relief by targeting the glutamatergic system in the brain. Ongoing research is exploring how ketamine works, its long-term efficacy, and its place in mental health treatment.

Psilocybin (Magic Mushrooms)

Psilocybin, the active compound in certain species of mushrooms, has garnered attention in recent years for its potential as a therapeutic treatment for depression and anxiety. Studies have shown that psilocybin can lead to lasting improvements in mood and cognition, with effects potentially lasting weeks or months after a single dose. Unlike SSRIs, which work gradually over time, psilocybin appears to have more immediate effects, leading some researchers to suggest it may be a promising alternative for certain individuals.

Neurostimulation Therapies

Transcranial magnetic stimulation (TMS) and deep brain stimulation (DBS) are non-invasive therapies that use electrical or magnetic fields to stimulate specific areas of the brain. These treatments are being explored as alternatives for patients who do not respond to medications like SSRIs. While they are still primarily used in research settings or for treatment-resistant cases, they represent exciting options for the future of mental health care.

Psychotherapeutic Advances

Alongside pharmacological innovations, psychotherapy is evolving as well. Cognitive-behavioral therapy (CBT), mindfulness-based therapy, and other therapeutic techniques are being integrated with pharmacological treatments to provide holistic and comprehensive care for patients. Future treatments may involve more collaboration between psychotherapists and psychiatrists, offering a balanced approach to mental health care.

3. The Integration of Digital Health Technologies

In addition to new medications and therapies, **digital health technologies** are playing an increasingly important role in the future of mental health treatment. Apps, online platforms, and virtual reality (VR) technologies are being developed to offer patients ways to manage their symptoms and track their mental health progress. These tools can supplement traditional treatments and provide patients with more autonomy in managing their well-being.

Mental Health Apps

There is a growing market of mental health apps that help individuals track their moods, manage stress, or even provide guided therapeutic exercises. Some apps offer cognitive-behavioral techniques, while others provide mindfulness or meditation practices. These apps could serve as adjuncts to medication, helping patients stay engaged in their treatment and providing valuable feedback to healthcare providers.

Virtual Reality (VR) Therapy

Virtual reality is being explored as a tool for immersive, exposure-based therapies. For example, VR can be used to simulate environments for patients with post-traumatic stress disorder (PTSD) or phobias, enabling them to confront their fears in a controlled, therapeutic setting. VR-based treatments are still in their infancy but show great potential in the future of mental health care.

4. The Role of Public Health Initiatives and Policy

As our understanding of mental health evolves, so too does the role of public health initiatives and policy. Increasing access to mental health care, reducing the stigma surrounding mental illness, and ensuring that treatments like SSRIs are available to all who need them will be critical components of future efforts to improve mental health outcomes globally. The integration of mental health into broader public health strategies—alongside innovations in therapy and treatment—will be a key factor in addressing the growing mental health crisis worldwide.

Conclusion: Moving Toward a Holistic Approach

The future of SSRIs and mental health treatment as a whole is an exciting one, driven by advances in pharmacology, psychotherapy, and technology. While SSRIs will likely remain a vital tool in managing conditions like depression and anxiety for the foreseeable future, new approaches and alternatives are continually being explored to enhance efficacy, improve patient outcomes, and ensure that treatments are personalized to individual needs.

Ultimately, the key to mastering serotonin reuptake inhibition—and mental health treatment more broadly—will be to view it as part of a holistic approach that combines medication, therapy, technology, and social support. By staying informed about emerging treatments and maintaining a proactive stance on mental wellness, patients can look forward to a future where mental health care is more effective, personalized, and accessible than ever before.

Chapter 19: The Role of Lifestyle Factors in Enhancing SSRI Effectiveness

While SSRIs (Selective Serotonin Reuptake Inhibitors) play a vital role in managing mental health conditions, such as depression, anxiety, and obsessive-compulsive disorder (OCD), they are only one part of a broader treatment strategy. Lifestyle factors—such as diet, exercise, sleep hygiene, stress management, and social support—can significantly influence the effectiveness of SSRIs and contribute to long-term mental health and emotional well-being.

In this chapter, we will explore how making positive changes in daily habits can help enhance the effects of SSRIs, mitigate side effects, and improve overall mental health outcomes. By integrating these lifestyle factors into a comprehensive treatment plan, individuals can experience greater benefits from their medication and better overall wellness.

1. Nutrition and Diet: Fueling Mental Health

A healthy, balanced diet is essential not only for physical health but also for mental well-being. Nutritional deficiencies, particularly in certain vitamins and minerals, can impact brain function and exacerbate symptoms of depression, anxiety, and other mood disorders. Recent research has highlighted the relationship between gut health and brain function, commonly referred to as the **gut-brain axis**, showing that what we eat can influence serotonin production, mood regulation, and cognitive health.

Key Nutrients for Mental Health

- **Omega-3 fatty acids** (found in fish, flaxseeds, and walnuts) are essential for brain health and have been shown to improve mood disorders, especially in conjunction with SSRIs.
- **B vitamins**, particularly B6, B12, and folate, are critical for serotonin production and regulation. Low levels of these vitamins can contribute to symptoms of depression.
- **Magnesium** plays a role in supporting brain function and has been shown to have a calming effect, reducing anxiety and promoting better sleep.
- **Probiotics** and **fiber-rich foods** (such as fruits, vegetables, and whole grains) are vital for gut health, which in turn supports serotonin production. About 90% of serotonin is produced in the gut, highlighting the connection between digestive health and mood regulation.

Dietary Considerations While on SSRIs

Certain foods and drinks, such as caffeine and alcohol, may interfere with SSRIs and exacerbate side effects like agitation, insomnia, or dizziness. Being mindful of these triggers and adopting a nutrient-rich diet can optimize SSRI effectiveness.

2. Exercise: The Natural Antidepressant

Exercise has long been recognized as a powerful tool in managing mental health. Physical activity boosts the release of endorphins, the body's natural "feel-good" hormones, and promotes the release of serotonin, helping to stabilize mood and improve emotional well-being. Regular exercise can complement the action of SSRIs by enhancing their mood-lifting effects and providing natural relief from stress, anxiety, and depression.

The Benefits of Exercise for Mental Health

- **Aerobic exercise**, such as walking, running, cycling, or swimming, is especially beneficial for improving mood and reducing symptoms of depression and anxiety.
- **Strength training** has been shown to have positive effects on mood regulation and cognitive function.
- **Yoga and mindfulness practices** provide physical exercise while also promoting relaxation and reducing stress.

For individuals on SSRIs, regular exercise can act as a natural antidepressant and may help mitigate some of the common side effects of SSRIs, such as weight gain and fatigue.

3. Sleep Hygiene: Restoring Balance to the Mind

Good quality sleep is essential for mental health. Unfortunately, sleep disturbances are common among those with mood disorders, and they can also be a side effect of SSRIs. While SSRIs can help regulate mood and improve sleep patterns over time, adopting healthy sleep habits is crucial for enhancing their effectiveness and ensuring long-term mental wellness.

Sleep Hygiene Tips

- Maintain a consistent sleep schedule, going to bed and waking up at the same time each day.
- Create a calming bedtime routine to signal to your body that it's time to wind down. This might include activities like reading, meditation, or taking a warm bath.
- Limit exposure to screens (phones, tablets, computers) at least an hour before bedtime to reduce the impact of blue light on your circadian rhythm.
- Avoid stimulants like caffeine or nicotine in the late afternoon and evening, as they can interfere with falling asleep.

Improving sleep quality can not only enhance the effectiveness of SSRIs but can also help improve energy levels, focus, and overall mood throughout the day.

4. Stress Management: Reducing the Impact of Daily Stressors

Chronic stress is one of the major contributors to mental health conditions such as depression and anxiety. While SSRIs help regulate mood, managing stress through relaxation techniques and coping strategies can further support mental health treatment. Learning how to reduce and cope with stress can enhance the benefits of medication and help maintain a positive outlook on life.

Effective Stress Management Techniques

- **Mindfulness meditation**: Practicing mindfulness and meditation can help regulate emotions, reduce anxiety, and improve overall mental well-being.

- **Deep breathing exercises**: Techniques like diaphragmatic breathing or the 4-7-8 method can activate the parasympathetic nervous system, helping to reduce stress and promote relaxation.

- **Progressive muscle relaxation (PMR)**: This involves systematically tensing and relaxing different muscle groups in the body to release physical tension and calm the mind.

- **Journaling**: Writing down thoughts and feelings can help process emotions, reduce rumination, and provide a sense of control over difficult situations.

Combining these techniques with SSRI treatment can create a more holistic approach to mental wellness, allowing individuals to manage both their physical and emotional responses to stress.

5. Social Support: Strengthening Emotional Connections

Having a strong support system is essential for mental health recovery. Connecting with others, whether through friends, family, or support groups, can provide encouragement, reduce feelings of isolation, and increase resilience in the face of challenges. Social support has been shown to enhance the effectiveness of SSRIs, as positive social interactions can increase serotonin levels naturally.

Building a Supportive Network

- Regularly engage with loved ones, share experiences, and maintain open communication about mental health.
- Consider joining a support group, either in person or online, to connect with others who are experiencing similar challenges.
- Seek professional help when needed, including therapy or counseling, to process difficult emotions and strengthen coping strategies.

Social support not only enhances the effects of SSRI treatment but can also provide the emotional tools necessary for navigating life's challenges.

Conclusion: A Comprehensive Approach to Mental Health

While SSRIs are an essential tool in the treatment of mood disorders, their effectiveness can be greatly enhanced when combined with healthy lifestyle practices. A balanced diet, regular exercise, good sleep hygiene, stress management, and strong social connections all contribute to a holistic approach to mental health. By addressing both the biological and environmental factors that influence mental well-being, individuals can experience more sustained recovery and improved overall quality of life. With this comprehensive approach, SSRIs can work more effectively, and long-term wellness can become an achievable and lasting reality.

Chapter 20: Building a Long-Term Plan for Wellness with SSRIs

Successfully managing mental health conditions often requires a long-term approach, and SSRIs (Selective Serotonin Reuptake Inhibitors) are just one tool in a comprehensive treatment plan. For individuals who have found relief through SSRI therapy, maintaining progress over time involves more than simply taking medication. Long-term wellness requires a balanced approach that combines ongoing medication management, psychological support, lifestyle adjustments, and self-care strategies. In this chapter, we will outline how to build a sustainable, long-term plan for mental health, integrating SSRIs as part of a holistic approach to well-being.

1. Ongoing Medication Management

SSRIs are often prescribed for long-term use, especially for chronic conditions like depression or anxiety. However, the landscape of mental health can shift over time...

Chapter 21: The Role of SSRIs in Preventing Relapse

One of the primary goals of SSRIs (Selective Serotonin Reuptake Inhibitors) in the treatment of mental health conditions is not only to alleviate symptoms but also to prevent relapse. For individuals recovering from depression, anxiety, or obsessive-compulsive disorder (OCD), the risk of symptoms returning after discontinuation or an adjustment in treatment is significant. SSRIs, when used correctly, can play a critical role in reducing the likelihood of relapse and supporting long-term recovery.

In this chapter, we will explore the ways in which SSRIs help prevent relapse, how long-term use can offer sustained benefits, and the strategies for integrating medication into a broader relapse prevention plan.

1. The Importance of Continued Medication Use

Although many individuals feel better after weeks or months of SSRI treatment and may be tempted to stop their medication, research shows that continued use of SSRIs—often at maintenance doses—can significantly reduce the risk of relapse. The brain's neurotransmitter systems, particularly serotonin, may take time to fully stabilize. Sudden discontinuation of SSRIs can lead to a recurrence of symptoms, even if an individual feels well.

Maintenance Treatment: How Long is Long Enough?

Studies show that for people who have experienced multiple episodes of depression, or for those who have had chronic anxiety, long-term or maintenance treatment with SSRIs can lower the chances of relapse by stabilizing serotonin levels in the brain. Generally, maintenance treatment lasts for at least 6-12 months following remission, but for those with a history of recurrent episodes, it may be beneficial to remain on medication for several years or even longer. Healthcare providers usually assess the risks and benefits of ongoing treatment regularly to determine the best course of action.

2. The Role of Therapeutic Support in Relapse Prevention

SSRIs are often most effective when combined with ongoing psychotherapy, particularly cognitive-behavioral therapy (CBT), which helps individuals build resilience and coping strategies for managing stressors and preventing future episodes. Cognitive-behavioral therapy focuses on changing negative thought patterns, developing problem-solving skills, and improving emotional regulation, all of which are essential for long-term mental health.

Relapse prevention is not just about medication, but about equipping individuals with the tools to manage their mental health without solely relying on pharmaceutical treatments. Therapy can provide a sustainable foundation for individuals to maintain their progress, even when reducing or discontinuing SSRIs.

3. Lifestyle Factors and Their Role in Preventing Relapse

Lifestyle changes and self-care routines are integral to preventing relapse. These factors complement SSRI treatment by addressing areas of physical and emotional well-being that directly influence mental health. Key lifestyle practices include:

- **Regular Exercise**: Physical activity has been shown to boost serotonin levels naturally and can help regulate mood and anxiety.

- **Balanced Diet**: Proper nutrition, particularly omega-3 fatty acids, vitamin D, and B-vitamins, can support brain health and reduce the risk of depression.

- **Stress Management**: Chronic stress is a major contributor to the onset and recurrence of mental health conditions. Practices like mindfulness meditation, yoga, or relaxation techniques can help manage stress effectively.

- **Adequate Sleep**: Poor sleep hygiene is linked to worsened mental health. Ensuring sufficient and restorative sleep is essential for emotional regulation.

Integrating these practices into daily life not only enhances the effectiveness of SSRIs but also strengthens mental resilience against potential future challenges.

4. Identifying Early Warning Signs of Relapse

An essential component of relapse prevention is recognizing the early warning signs that symptoms may be returning. These may include:

- A return of negative thinking or rumination

- Increased irritability or anxiety

- Disrupted sleep patterns or insomnia

- Loss of interest in activities that were once enjoyable

- Changes in appetite or energy levels

Early intervention is key to preventing a full relapse. Individuals on SSRIs should be encouraged to remain vigilant for these signs and to seek help from a healthcare provider if symptoms begin to resurface. Monitoring progress with a therapist or counselor, as well as regularly reviewing treatment plans, can help catch potential relapses early.

5. Gradual Reduction and Tapering Off SSRIs

For some individuals, long-term SSRI treatment may no longer be necessary once they have achieved long-term stability. However, discontinuing medication must be done gradually and under the guidance of a healthcare provider. Tapering off SSRIs slowly reduces the risk of withdrawal symptoms and allows the brain to adjust to the change.

While some people may eventually discontinue SSRI treatment without a relapse, this decision should be made collaboratively with a healthcare provider, considering the individual's medical history, mental health condition, and long-term goals.

Conclusion

SSRIs have proven to be effective in preventing relapse and maintaining long-term stability for many individuals with mental health conditions. When used as part of a comprehensive treatment plan that includes ongoing medication, psychotherapy, lifestyle changes, and vigilant monitoring for warning signs, SSRIs can help reduce the risk of relapse and support sustained emotional well-being.

A proactive approach to relapse prevention not only increases the likelihood of successful long-term recovery but also empowers individuals to take charge of their mental health, fostering greater resilience and a better quality of life. By integrating all these strategies, individuals can build a robust framework for maintaining mental wellness, with SSRIs as a vital part of the journey toward lasting stability and happiness.

Chapter 22: Ethical Considerations and the Future of SSRI Prescribing

The use of SSRIs (Selective Serotonin Reuptake Inhibitors) has significantly changed the landscape of mental health treatment. However, with their widespread use come a variety of ethical considerations that healthcare providers must address to ensure that patients are receiving the most appropriate care. This chapter will explore the ethical challenges surrounding the prescribing of SSRIs, such as balancing the risks and benefits, the potential for over-prescribing, the importance of informed consent, and the need for individualized treatment plans. As the field of psychiatry evolves, these ethical questions will remain relevant in guiding the responsible use of SSRIs in the treatment of mental health disorders.

1. Balancing Benefits and Risks: Ethical Dilemmas in SSRI Use

While SSRIs are widely regarded as a safe and effective treatment for various mental health conditions, they are not without risks. Understanding the ethical responsibility of prescribing these medications involves considering the balance between their therapeutic benefits and the potential side effects. Healthcare providers must assess whether the benefits outweigh the risks for each patient, factoring in both the severity of the disorder and the individual's medical history.

For example, SSRIs can be highly effective in treating depression and anxiety, but they also come with a risk of side effects, ranging from mild (such as headaches or nausea) to more serious (including suicidal thoughts, particularly in younger patients). Ethical prescribing requires careful consideration of these risks and transparent communication with patients about what they can expect from treatment. Providers should assess whether other treatment options, such as therapy or lifestyle changes, may be better suited for patients who are at higher risk for side effects or who have a history of adverse reactions to medications.

2. Informed Consent and Patient Autonomy

A fundamental ethical principle in healthcare is the importance of informed consent, which ensures that patients are fully aware of their treatment options, the expected benefits, and the potential risks before agreeing to a course of action. With SSRIs, informed consent includes educating patients about the possible side effects, the time it may take for the medication to start working, and the importance of adhering to the prescribed dosage.

Furthermore, informed consent requires that patients are aware of their right to discontinue the medication if they feel that it is not working or if they experience intolerable side effects. Patients should also be informed about the risks of withdrawal symptoms when discontinuing SSRIs, and the importance of tapering under medical supervision.

In some cases, SSRIs are prescribed to individuals who are hesitant about taking medication, or who may have concerns based on previous experiences with mental health treatment. Ensuring that patients understand the rationale for using SSRIs, and allowing them to express any reservations, helps foster a relationship of trust and respect between the patient and their healthcare provider. This approach is essential for maintaining patient autonomy and ensuring that the patient's voice is heard throughout the treatment process.

3. The Risk of Over-Prescribing SSRIs

Over-prescribing of SSRIs is another important ethical issue in modern psychiatric care. The growing trend of prescribing antidepressants in response to mental health symptoms has raised concerns about the potential overuse of medication, especially in cases where alternative treatments, such as psychotherapy or lifestyle modifications, might be just as effective—or more appropriate.

Over-prescribing can also occur when SSRIs are prescribed for conditions that are not well-supported by evidence, such as mild anxiety or transient mood changes, which may be better managed with non-pharmacological interventions. There is also the risk of relying on SSRIs as a "quick fix," rather than addressing underlying issues through holistic treatments that include therapy, social support, and lifestyle changes.

Ethical prescribing practices require healthcare providers to carefully evaluate whether SSRI treatment is necessary and appropriate, and whether other treatment options may be more beneficial. In some cases, SSRIs should be viewed as part of a comprehensive treatment plan that includes psychotherapy, support networks, and lifestyle interventions.

4. Personalized Treatment Plans: Tailoring Care to the Individual

The principle of individualized care is another key ethical consideration in SSRI prescribing. Mental health treatment is not a one-size-fits-all approach, and SSRIs should be prescribed in a way that is tailored to each patient's specific needs and circumstances. Factors such as age, comorbid conditions, prior medication history, family dynamics, and even personal preferences should all play a role in developing a treatment plan.

Personalized care means taking a more holistic approach to treating mental health conditions, incorporating not just medication, but also therapy, lifestyle modifications, and social support. It also involves continuously monitoring the patient's progress and adjusting the treatment plan as necessary. Ethical prescribing requires flexibility and a willingness to adapt as new information emerges about the patient's condition and treatment response.

5. The Future of SSRI Prescribing: Emerging Trends and Challenges

As the field of psychiatry continues to evolve, so too will the ethical considerations around SSRI prescribing. New research into the long-term effects of antidepressants, as well as alternative treatments such as psychedelic-assisted therapy or neuromodulation, may challenge the current standard of care. The future of SSRI prescribing will likely involve more personalized approaches, leveraging advances in genetics, pharmacogenomics, and neuroimaging to understand which treatments are most effective for individual patients.

As we look toward the future, it is important that we continue to address these ethical challenges, ensuring that SSRIs and other treatments are used responsibly and effectively. Ethical prescribing of SSRIs will always prioritize patient well-being, informed consent, and the promotion of holistic, individualized care.

Conclusion

The responsible and ethical use of SSRIs in treating mental health conditions requires a delicate balance between benefits, risks, and patient autonomy. Healthcare providers must navigate the complexities of prescribing SSRIs, ensuring that each patient receives the most appropriate and effective treatment for their unique needs. By prioritizing informed consent, minimizing over-prescribing, and emphasizing individualized care, the future of SSRIs in mental health treatment can be both effective and ethically sound.

Chapter 23: The Global Impact of SSRIs on Mental Health Treatment

Mental health issues are a global concern, affecting millions of people worldwide. As research into the biological and psychological factors that contribute to conditions like depression, anxiety, and PTSD continues to advance, SSRIs (Selective Serotonin Reuptake Inhibitors) have become one of the most widely prescribed classes of medication for managing mood disorders. The global spread of SSRIs has not only changed the way we approach mental health treatment but has also highlighted disparities in access, care, and cultural attitudes toward mental health.

This chapter will explore the global impact of SSRIs on mental health treatment, examining their availability and usage in different regions, the role of cultural perceptions in medication adherence, and the challenges of mental health care in underserved areas.

1. The Widespread Use of SSRIs in Developed Countries

In high-income countries, SSRIs have become a staple of mental health treatment. From the United States and Canada to Europe and Australia, millions of people rely on SSRIs to manage a variety of mental health conditions. The convenience and efficacy of SSRIs, combined with the growing recognition of mental health as an important part of overall wellness, have helped reduce the stigma surrounding psychiatric medication in many parts of the world.

However, despite their widespread use, there are still challenges. For instance, **long-term adherence to treatment** is a significant issue. Patients may discontinue SSRIs prematurely due to side effects, lack of immediate improvement, or feelings of dependency. This chapter will look at efforts in these regions to improve adherence, including public health campaigns, mental health awareness, and evolving healthcare models that emphasize integrated care.

2. Challenges of Access and Use in Low- and Middle-Income Countries

While SSRIs are readily available in many developed countries, **access to SSRIs** remains a major challenge in low- and middle-income countries (LMICs). Cost, limited healthcare infrastructure, and a lack of trained mental health professionals can all contribute to difficulties in accessing proper mental health treatment. In some countries, psychiatric medications are viewed with skepticism, or the use of medications like SSRIs is limited to the wealthier segments of the population.

In regions with limited mental health care, **traditional healing methods** and community-based approaches often serve as primary sources of care. These practices may not always align with evidence-based medical treatments but play an important cultural role in managing mental health. As a result, there is often a gap between available medical treatment and the healthcare preferences of the population.

Efforts are being made globally to bridge this gap by **lowering the cost of SSRIs**, improving access to medication, and promoting mental health education. This section will highlight the challenges and the innovative solutions being implemented to address these issues, such as **telemedicine**, **mobile health apps**, and **psychiatry training programs** for primary care providers.

3. Cultural Perceptions of Mental Health and SSRIs

Cultural attitudes toward mental health vary widely around the world. In some cultures, mental health conditions are seen as personal weaknesses or stigmatized as moral failings, which can discourage individuals from seeking help. These cultural perceptions can have a profound impact on the **acceptance of SSRIs** and other pharmacological treatments. While antidepressant medications like SSRIs have become widely accepted in many parts of the world, in other areas, there is resistance to using medication, especially for conditions like depression or anxiety.

The **stigma surrounding mental health** treatment is a key barrier to increasing global access to SSRIs. In some cultures, there is a preference for non-medical treatments, such as **therapy, spirituality, or social support**. However, in countries where SSRIs are more commonly prescribed, public attitudes are changing, particularly as more information becomes available about the biological underpinnings of mental health disorders.

Efforts to increase **global mental health literacy** are essential to shifting these perceptions and ensuring that people who need medication have access to it. Educational campaigns and advocacy work are helping to normalize mental health treatment, reduce stigma, and encourage more people to seek help when needed.

4. The Future of Global Mental Health and SSRIs

As the global conversation around mental health continues to evolve, SSRIs will likely remain a central part of the treatment landscape. However, there is increasing interest in finding new and improved treatments that offer better outcomes, fewer side effects, and greater accessibility. Advances in **personalized medicine**, **genetic testing**, and **neurobiological research** will likely shape the future of SSRI prescribing.

In addition to pharmaceutical innovations, there is a growing recognition of the importance of a **holistic approach** to mental health, integrating **psychotherapy**, **lifestyle changes**, **community support**, and **technology** alongside traditional pharmacological treatments. As these different strategies come together, the global mental health landscape will continue to change, ideally improving the well-being of individuals across the globe.

5. Conclusion

The global impact of SSRIs on mental health care has been profound, helping millions of people manage mental health conditions and improve their quality of life. However, there are still significant barriers to access, cultural attitudes to address, and new challenges to overcome in the effort to improve mental health treatment worldwide. As research continues, it is essential to balance the widespread use of SSRIs with a global effort to provide equitable, effective, and culturally sensitive care for all individuals, regardless of where they live.

This chapter has highlighted some of the challenges and successes in the global use of SSRIs, and it serves as a reminder that the future of mental health care is not only about better medications but also about breaking down the barriers to access, reducing stigma, and promoting a more inclusive approach to mental well-being. The continued evolution of SSRIs and mental health treatment will be shaped by the collective efforts of healthcare providers, patients, communities, and global health organizations.

Chapter 24: Integrating SSRIs into a Holistic Mental Health Approach

While SSRIs (Selective Serotonin Reuptake Inhibitors) are effective medications for managing mental health conditions like depression, anxiety, and obsessive-compulsive disorder (OCD), their optimal effectiveness often depends on their integration into a broader, more holistic mental health treatment plan. The most successful mental health strategies combine medication, therapy, lifestyle changes, and ongoing support systems. This chapter will explore how to effectively integrate SSRIs into a comprehensive approach to mental well-being that addresses the mind, body, and environment, ensuring long-term success and improved quality of life.

1. The Role of Psychotherapy and SSRIs

Although SSRIs can help alleviate symptoms of mood disorders by balancing serotonin levels, they are most effective when combined with psychotherapy. Cognitive-behavioral therapy (CBT), mindfulness-based therapy, and other therapeutic approaches provide individuals with the tools to manage thoughts, emotions, and behaviors that contribute to mental health challenges. When used together, medication and therapy can complement each other and promote lasting healing.

- **Cognitive Behavioral Therapy (CBT)**: CBT focuses on changing unhelpful thinking patterns that contribute to mental health symptoms. Research has shown that combining SSRIs with CBT can improve treatment outcomes for depression and anxiety by addressing the underlying cognitive distortions while SSRIs provide chemical stabilization.

- **Mindfulness-Based Stress Reduction (MBSR)**: Mindfulness practices, often incorporated into therapy, help individuals regulate emotions, reduce stress, and enhance resilience. Research indicates that combining SSRIs with mindfulness practices can further reduce anxiety and improve emotional regulation.

2. Lifestyle Modifications to Support Mental Health

SSRIs work by altering serotonin levels in the brain, but they cannot address all the factors that influence mental well-being. Lifestyle modifications such as regular exercise, adequate sleep, a healthy diet, and stress reduction techniques are critical in enhancing the benefits of SSRIs.

- **Exercise**: Physical activity has been shown to increase serotonin levels naturally, improving mood and reducing anxiety and depression. Engaging in regular exercise not only boosts the effects of SSRIs but also contributes to overall well-being, helping to combat the lethargy or low energy that can sometimes accompany mental health challenges.
- **Diet and Nutrition**: A balanced diet rich in omega-3 fatty acids, antioxidants, and vitamins can support brain function and improve mood regulation. Incorporating more whole foods, such as fruits, vegetables, and lean proteins, while avoiding processed foods, can help individuals feel better physically and emotionally.
- **Sleep Hygiene**: Poor sleep is often a symptom of depression and anxiety, and it can also make these conditions worse. Ensuring adequate and restful sleep through proper sleep hygiene (e.g., regular sleep schedule, limiting screen time, and creating a relaxing environment) is vital in supporting both mental health and the effectiveness of SSRIs.

3. Building a Support System

Social support plays a key role in mental health recovery. Whether it comes from family, friends, support groups, or mental health professionals, having a reliable network is crucial for managing stress, coping with setbacks, and maintaining motivation. Studies show that people who have a strong social network are more likely to experience improved outcomes in mental health treatment.

- **Family and Friends**: A supportive home environment, where open communication is encouraged, can help individuals feel understood and empowered to engage in treatment. Family members can also play a role in monitoring for potential side effects or signs of relapse.
- **Support Groups**: Peer support groups, whether in-person or online, offer a space for individuals with similar experiences to share strategies, challenges, and successes. These groups often provide valuable social interaction and reduce feelings of isolation.
- **Professional Support**: Alongside SSRIs, maintaining regular appointments with a mental health professional, such as a psychiatrist, psychologist, or therapist, is essential for ongoing management. These professionals can help patients navigate the ups and downs of treatment, adjust medication if necessary, and offer guidance during challenging periods.

4. The Importance of Self-Care and Stress Management

Self-care practices are essential in maintaining mental health and supporting the healing process. Stress, when left unchecked, can undo the progress made through medication and therapy. Incorporating daily stress management strategies—such as deep breathing, yoga, journaling, or creative hobbies—can help reduce the impact of life's challenges on mental health.

- **Mindfulness and Meditation**: These practices can help individuals manage stress by encouraging present-moment awareness and reducing rumination, a common issue in depression and anxiety. Regular meditation has been shown to complement the effects of SSRIs by improving emotional regulation and resilience.
- **Relaxation Techniques**: Techniques such as progressive muscle relaxation, guided imagery, and aromatherapy can help reduce physical symptoms of stress, like muscle tension and rapid heart rate, thereby promoting a sense of calm.

5. Monitoring Progress and Adjusting Treatment

Mental health is dynamic, and treatment plans must be flexible to meet changing needs. Regular check-ins with healthcare providers—whether through medication adjustments, changes in therapy approaches, or updates in lifestyle strategies—are necessary for maintaining progress and preventing relapse.

- **Medication Adjustments**: SSRIs are effective for many individuals, but some may need adjustments to their dosage or medication regimen over time. For instance, a patient may find that a particular SSRI becomes less effective or that side effects become problematic, requiring a shift to another medication.

- **Therapy Adjustments**: As individuals progress in their recovery, their therapeutic needs may evolve. Therapy goals may shift, or new techniques may be introduced to address emerging challenges. Ensuring that therapy remains relevant to the individual's current state is vital for long-term success.

- **Lifestyle Adjustments**: As personal circumstances change—such as changes in work, relationships, or physical health—adjustments to exercise routines, diet, and sleep habits may be necessary to maintain mental health and the effectiveness of SSRIs.

6. Creating a Sustainable Treatment Plan

Ultimately, building a sustainable treatment plan for mental health involves creating a balanced, adaptable framework that incorporates SSRIs, therapy, healthy lifestyle choices, and ongoing support. The goal is to develop an approach that not only addresses symptoms but fosters long-term emotional resilience, well-being, and self-efficacy. It is a comprehensive, dynamic process that supports individuals in living fulfilling, healthy lives despite their mental health challenges.

Conclusion

Integrating SSRIs into a holistic mental health treatment plan is not about relying on medication alone, but about creating an environment of care that nurtures mind, body, and spirit. By combining medication, therapy, lifestyle changes, and social support, individuals can maximize the benefits of SSRIs, enhance their mental well-being, and build resilience for the future. The key to lasting mental health is a comprehensive, integrated approach that empowers individuals to take control of their health and lead fulfilling lives.

Chapter 25: Conclusion: Mastering Mental Health with SSRIs and Beyond

The journey of understanding and utilizing SSRIs (Selective Serotonin Reuptake Inhibitors) as a key tool in managing mental health challenges is one that is multi-faceted and evolving. From their discovery and the groundbreaking role they play in treating conditions like depression, anxiety, and OCD, to the ongoing research into their effectiveness and potential, SSRIs have proven to be an indispensable part of modern mental health care. However, as we've seen throughout this book, SSRIs are most effective when integrated into a holistic approach to mental wellness, one that includes lifestyle changes, therapy, social support, and a proactive mindset.

In this final chapter, we reflect on the critical takeaways from this comprehensive guide and consider how individuals can continue to benefit from SSRIs while also optimizing their mental health through other means. By understanding the role of serotonin, embracing the safety and side effects of SSRIs, and committing to a broader wellness plan, patients can not only manage their symptoms but thrive in the long-term.

1. The Ongoing Journey of Mental Wellness

Managing mental health is not a one-time fix, but a lifelong journey. SSRIs provide an essential tool in this journey, but they work best as part of a broader strategy that includes mental health education, self-care, physical wellness, and ongoing therapy. While SSRIs can significantly reduce the symptoms of mental health conditions, they do not necessarily address all underlying factors. This is why the holistic approach discussed throughout this book remains essential.

For individuals taking SSRIs, the key to success is continuous self-monitoring, adapting the treatment plan as needed, and maintaining an open dialogue with healthcare providers. It's important to remember that there are many paths to mental well-being, and SSRIs are only one of those paths.

2. Looking Ahead: Innovations and the Future of Mental Health Treatment

The future of mental health care is filled with possibilities. While SSRIs remain a first-line treatment for many, research continues to uncover new ways of treating mood disorders. Advances in neurobiology, gene therapy, and personalized medicine may offer new, more targeted treatments that address the root causes of mental health conditions in ways SSRIs do not. As scientific understanding grows, SSRIs themselves may evolve, becoming even more effective with fewer side effects.

Moreover, as we continue to learn more about the mind-body connection, mental health care will likely place greater emphasis on integrating physical health, social support, and environmental factors in comprehensive treatment plans. This shift toward holistic care will empower individuals to not only manage their symptoms but also foster a deeper understanding of their mental health.

3. Empowering the Individual: Taking Control of Your Mental Health

Ultimately, the power to improve mental health lies in the hands of the individual. SSRIs can provide support in regulating mood and managing mental health conditions, but it is the individual's active participation in their recovery that leads to long-term success. This includes taking ownership of their treatment plan, cultivating healthy lifestyle habits, practicing mindfulness, seeking therapy, and building strong social connections. Empowering oneself through education, self-awareness, and resilience is key to mastering mental health and achieving overall well-being.

4. Final Thoughts: A Balanced Approach to Wellness

While SSRIs have revolutionized mental health treatment and continue to play a crucial role in the management of various disorders, true mental wellness comes from a balanced approach that includes the right medication, emotional support, self-care, and personal growth. As we move forward, embracing a multifaceted strategy for mental health is essential, and SSRIs will continue to be an important tool in that strategy.

By understanding how SSRIs work, acknowledging their potential side effects, and integrating them into a comprehensive wellness plan, individuals can achieve not only symptom relief but also long-term mental health and emotional resilience.

www.ingramcontent.com/pod-product-compliance
Lightning Source LLC
Chambersburg PA
CBHW082107220526
45472CB00009B/2082